BEAUTIFUL
A Celebration of Evolution

William Gurney

BPP

For my beloved Katia, for everything, and for Neil for his unwavering support and belief. – W.S.

BIG PICTURE PRESS

First published in the UK in 2024 by Big Picture Press,
an imprint of Bonnier Books UK,
4th Floor, Victoria House
Bloomsbury Square, London WC1B 4DA
Owned by Bonnier Books
Sveavägen 56, Stockholm, Sweden
www.bonnierbooks.co.uk

Text and illustration copyright © 2024 by William Spring
Design copyright © 2024 by Big Picture Press

1 3 5 7 9 10 8 6 4 2

All rights reserved

ISBN 978-1-80078-616-5

This book was typeset in Le Havre Sans and Bodoni
The illustrations were created in pencil and watercolour

Edited by Charlie Wilson
Designed by Winsome d'Abreu and Jenny Hastings
Production by Ché Creasey

Printed in China

Contents

Our Beautiful World	6
Aldabra Giant Tortoise	8
American Bison	10
Andean Condor	12
Aye-Aye	14
Bat-Eared Fox	16
Camel	18
Chambered Nautilus	20
Chameleon	22
Duck-Billed Golden-Line Barbel	24
Duck-Billed Platypus	26
Electric Eel	28
European Mole	30
Fangtooth Fish	32
Fossa	34
Giant Anteater	36
Helmeted Guinea Fowl	38
Hippopotamus	40
Hornbill	42
Horsehoe Crab	44
Ibis	46
Indian Elephant	48
Indian Pangolin	50
Malayan Tapir	52
Naked Mole Rat	54
Narwhal	56
North Sulawesi Babirusa	58
Northern Three-Toed Jerboa	60
Ocean Sunfish	62
Orange Oakleaf Butterfly	64
Pink Fairy Armadillo	66
Portuguese Man-of-War	68
Pygmy Seahorse	70
Saiga Antelope	72
Saltwater Crocodile	74
Scottish Red Deer	76
Shoebill	78
Southern Cassowary	80
Sperm Whale	82
Spider Monkey	84
Spotted Bat	86
Spotted Hyena	88
Striped Skunk	90
Sumatran Rhino	92
Sunda Colugo	94
Toad	96
Treehopper	98
Vampire Squid	100
West African Giraffe	102
White-Faced Saki Monkey	104
Wild Boar	106
Glossary	108
About the Artist	110

Our Beautiful World

The natural world, and an appreciation of it, is the greatest gift humans can experience – from the tiniest insect or pollen grain to the largest mammal or the tallest tree.

But too often we label many of the animals who inhabit our world as 'ugly'. This way of thinking disregards the complexity of nature and the vital role all creatures play in the fine balance that is our ecosystem. It also fails to acknowledge the process of evolution, and how each and every animal has adapted to its environment.

Every single living thing on this planet is the product of evolution by natural selection. This means that animals evolve over time to adapt to their environment and aid their survival within it. As living things reproduce, occasionally an animal will be born with a mutation, which is a very slight change in its genetic make-up. This mutation might be useful for the animal and help it survive. If this is the case, then that animal will have a higher chance of breeding and passing that same useful feature to its offspring. Over generations, that small change can become more pronounced and more useful to the animal.

A good example is the evolution of the eye. Imagine, back in the mists of time, an animal without eyes. However, this animal has been born with a mutation – it has facial cells that are sensitive to light and dark. When a predator comes towards it and casts a shadow, our animal detects this change in light and darts away, saving its life. Because it has survived, it can breed. Its offspring inherit these light-sensitive cells, which also help them survive. The process continues until eventually the entire species can detect light. These light-detecting cells gradually change position to be set slightly lower in the skin, forming a pit that helps focus the light, giving the animal an even greater ability to detect changes. Over time, these pits develop, becoming deeper and more enclosed. A membrane covering the pit forms a lens, and so on and so forth until the animal has a pair of highly developed eyes.

This process of developing tiny changes over enormous stretches of time applies to every single aspect of every single living thing, from the smallest single-celled amoeba to the largest animal on the planet – the blue whale. An animal's appearance and the way it has evolved are, broadly speaking, a result of its adaptations for survival, aiding them in feeding, avoiding predators and successfully reproducing.

The truth of the matter, and the point I hope this book will illustrate, is that in nature the word *ugly* simply does not apply. All living things have adapted to their environment in unique and often startling ways. Bacteria, trees, whales, you and I and everything else alive today have all descended from a common ancestor billions of years ago. It's an unbroken chain of survival. To destroy an animal because of its appearance, because of fear or (as is more common) simply because it is an

inconvenience, is to destroy the end result of billions of years of evolution.

Humans think in short spans of time. Our lifetimes last for a tiny fraction of the history of the Earth, so it's hard for us to think in spans of thousands or millions of years. But even these longer stretches of time are reasonably small — the Earth, along with the life on it, has existed for billions of years. It will continue to exist for billions more. If we can influence our own evolution, then the best we can do for our own survival is for all of us to adapt our behaviour. We need to realise that we ourselves are the greatest threat that nature currently faces. We must become caretakers of this world, protecting and nurturing all living things. Our own survival depends on it.

Evolution makes all of nature indescribably beautiful. Everything in nature is a treasure that each and every one of us must appreciate and protect. This book contains a sample of the wonders inhabiting this planet. They are all beautiful.

Aldabrachelys gigantea

Aldabra Giant Tortoise

It was the tortoise from Aesop's famous fable The Hare and the Tortoise *who gave us the phrase 'slow and steady wins the race'. It couldn't be truer for this animal, especially when put into context. Tortoises have been gently plodding around the planet for 300 million years.*

This gentle, unassuming creature is one of the longest-lived animals in the world. In the wild, a healthy tortoise can live to be more than 150 years old. Some have even been recorded as living to 250 years. The reason that these animals live for so long is somewhat contested. One theory claims that it's because their metabolism – the process of burning energy – is slow. Another theory is to do with their habitat. Tortoises evolved in isolated areas with few natural predators. Perhaps they simply didn't need to evolve a fast reproductive cycle and the ability to exit quickly when danger appeared.

One of the tortoise's most well-known traits is its ability to withdraw inside its shell – or, more accurately, its carapace. Contrary to popular belief, the shell is not one solid piece of bone. It is actually made of many interlocking plates called scutes. These are made of keratin, the same substance that makes human fingernails and rhinoceros horns (see page 92). The carapace is also extremely sensitive. It is filled with nerve endings that allow the tortoise to feel even the lightest touch – an important ability when you're hiding inside your home and you need to know when a predator has stopped trying to bite it open! The bottom part that covers the belly is called the plastron, and the top and bottom halves are connected by a flexible section called the bridge. These three sections are not rigid, allowing the tortoise the flexibility to move around inside its shell. Of course, within its body the tortoise has a skeleton made of bone. However, unlike other vertebrates, it has a completely fused and inflexible spine, meaning that it is unable to move its back. As a result, it has evolved a long, immensely strong and flexible neck to reach food, such as leaves, that would otherwise be inaccessible.

The Aldabra giant tortoise pictured here (one of the largest species, often measuring more than a metre in length and height) is one of the few species that thrive today, mainly because of its unusually long conservation history. In the eighteenth century, hungry European sailors would stack tortoises aboard their ships like barrels. Easy to catch and able to survive a great deal of time without food or water, tortoises were the perfect food. While many species were

driven to extinction by overexploitation, the Aldabra giant tortoise, found on a small island in the Seychelles, was fortunately recognised as being endangered at a very early stage, and was enrolled in an early conservation project by zoologist Albert Günther, who worked with the Mauritian government.

    Today, the Aldabra giant tortoise is protected from humans and invasive species, resulting in an estimated population of one hundred thousand. Let's hope that they are all living out their long lives in peace.

Bison bison

American Bison

The formidable American bison, the largest mammal in North America today, sadly illustrates an often-told and tragic story of humans bringing an animal to the brink of extinction. By definition, this also brings enormous damage to an ecosystem.

The bison is an incredible beast. Sometimes referred to as the buffalo (a word that is believed to come from *boeuf*, the French word for beef), an adult bison weighs up to 1,100 kilograms and stands at more than 2 metres tall. Capable of running at a speed of almost 64 kilometres per hour, it can also rapidly spin and pivot, as well as jump its full body height into the air. As if this wasn't remarkable enough, its great head, which the male uses in combat to fight for a mate, has an incredibly thick skull and two curved horns.

Life out on the open plains, where animals are exposed to the elements and freezing conditions in winter, has led to the bison developing two of its most recognisable features — its enormous shoulder hump and its distinct shaggy fur. When ancestors of today's American bison were faced with heavy prolonged snowfall, the huge hump, made almost entirely from muscle, allowed them to push through extremely deep snow drifts, clearing a path and revealing the hidden grasses below.

Unlike other animals, bison do not burn valuable calories to stay warm. Instead, they have an incredibly heavy and specialised coat to protect them from the harsh winters they endure. This coat comprises of two distinct layers. The inner layer is made up of lots of very fine hairs that trap air, and therefore heat. A coarser outer layer of hair acts almost like a jacket, protecting the animal from the winds and snow. The bison's coat is so efficient that these animals are often seen covered in snow, which does not melt when it settles on their fur.

Herbivores are typically very large animals, and the bison is no exception. As they consume plants directly from the source, there is no need for them to chase after prey, and the vast areas they roam allow them to eat enough to grow to an immense size. Like all animals, bison fill a very important role in the ecosystem, grazing on the native grasses of the plains and using their hooves to aerate the soil and their droppings to fertilise it. The impact bison have on their environment has also influenced other species that live alongside them.

But not every species has managed to live in harmony with them. During the nineteenth century, European settlers slaughtered more than 30 million bison in a matter of a few decades — and that is a conservative number. They were killed for food, sport and business (in 1871, it was discovered that bison hides could be used for industrial leather), and, sadly, to deprive the local Native Americans of their most valuable resource. As a result, bison were brought to near total extinction, with just a few hundred animals remaining from pre-hunting estimates of 60 million. Today, numbers of wild populations are still incredibly low, at just 25,000–30,000 individuals. Existing herds have relatively few members, making genetic diversity a concern for survival.

Vultur gryphus

Andean Condor

The breathtaking Andean condor has to be one of the most spectacular and impressive birds on Earth.

With a wingspan of over 3 metres and an impressive bulk of around 13.5 kilograms, the Andean condor is the largest flying bird in the world, and a truly staggering sight in the wild. Living in the high mountains of the Andes and along the Pacific coast of South America, the Andean condor spends its days gliding over the landscape looking for food. It is a scavenger, feeding on the bodies of dead animals, which are often hard to come by. Because of this, the Andean condor has evolved enormous wings to catch air currents and conserve energy, allowing it to glide over a distance of 160 kilometres without flapping its wings once, at altitudes of more than 5,000 metres above ground.

You may think that flying at such heights would make it tricky for the condor to spot food, but this is far from being the case. Like many birds of prey, the Andean condor has developed acute eyesight, allowing it not only to spot a carcass from several kilometres away but also to watch the behaviour of groups of other scavengers that might be gathering around an animal that has recently died.

Another useful adaptation, and one shared by the southern cassowary (see page 80), is its small, bald head. Rotting carcasses are host to a huge variety of bacteria, parasites and maggots, and an animal feeding on such a diet needs to avoid illness. As well as developing a high tolerance to various bacteria, the condor has lost the feathers on its head, allowing it to feed more hygienically. In fact, it is so particular about keeping clean after a meal that it has been observed rubbing its head along the ground, using the grass and rocks like a napkin. Humans have a reason to be thankful to the Andean condor and its feeding habits. As carrion eaters, these condors play a vital role in the environment, disposing of dead animals that would otherwise potentially cause the spread of disease.

As its food is often quite scarce, the Andean condor will gorge itself when possible, feeding until it is so heavy it cannot immediately lift off again and fly. This might be a problem for smaller birds, but the condor's great size and sparsely populated environment means it has few (if any) natural predators, so can digest its meal in peace.

Although this incredible bird has evolved without any natural predators and has a naturally high rate of survival, it has sadly not been able to adapt quickly enough to the impact of humankind. Habitat destruction is a particular threat, as well as being hunted – the condor is believed to threaten livestock, despite not having talons or a well-suited beak. This magnificent bird is now considered vulnerable, and in some areas it is already virtually extinct. This wonder of nature is under enormous threat of disappearing forever, and efforts are under way to protect the world's largest flying bird.

Daubentonia madagascariensis

Aye-Aye

T*he elusive aye-aye is a marvel of nature. Nocturnal, solitary and living almost its entire life within the canopy of the Madagascan rainforests, it has developed some marvellous adaptations to help it locate its food and live life in the dark.*

Huge forward-facing orange eyes allow this little lemur to see, while its enormous bat-like ears have evolved to be extraordinarily sensitive, containing a complex set of ridges and grooves which help detect the tiny sounds coming from the creatures it feeds on. This is an adaptation shared with other animals that listen for their food, such as the spotted bat (see page 86).

Yet the aye-aye's most well-known feature — and perhaps its most useful — is its elongated middle finger. To find its prey (mostly grubs and small insects), the aye-aye uses this special digit to rapidly tap along the branch, like a woodpecker, listening with its huge ears for the sound of a grub moving beneath the bark. This technique is called percussive foraging. Once the aye-aye has located a tasty grub, it uses its strong incisors to bite and chew its way through the wood to reveal the tunnel the grub has made. Here, its incredible finger is used again. Long and bony, it is slender enough to fit within the narrow tunnels, and its sharp claw is perfect for hooking out and retrieving the unsuspecting grub. Additionally, this spindly finger comes in handy for scooping out pulp from the tasty fruits it also eats.

Owing to the rapid destruction of its habitat in Madagascar, the aye-aye is now critically endangered. Worst still, these mysterious creatures are often considered bad omens by the Malagasy. According to local legend, aye-ayes signify bad luck or death. Unfortunately, this means these animals are often killed on sight.

Successful captive breeding programmes could turn round the fate of this animal, but careful conservation of its natural habitat and a 'rebranding' of the aye-aye among the local community is essential if this unique creature is to survive.

Otocyon megalotis

Bat-Eared Fox

The name explains it all: a fox with ears like a bat. But why has this extraordinary animal developed such unusual features?

Living in small, tightly knit social groups, these foxes make their home on the savannahs of eastern and southern Africa. This is a dry and hot landscape where food is limited and danger is ever-present. The fox's large ears are an adaptation shared by many other inhabitants of these kinds of environments, such as the fennec fox, showing how evolution works to find the most efficient 'solution' to the problem of survival (bearing in mind that nothing is ever 'finished' in evolution — it's a constant work in progress).

In the case of our fox, its specialised ears serve several purposes. Firstly, they give it an astonishing sense of hearing. Not only does this allow the fox to detect predators and keep itself safe, but it also helps it to be a predator itself. It consumes a wide range of prey, including moths, spiders and millipedes, and occasionally even small reptiles and mammals — all of which are elusive, fast and, more often than not, actually hidden, as they live or shelter underground. In such situations, the ability to hear your dinner before you see it makes it a lot easier to locate. Bat-eared foxes can often be seen walking slowly with their noses to the ground (their sense of smell is also extremely sensitive), their huge ears cocked forwards, listening for the tiny scratch of an insect in its burrow. Once the fox has located its prey, it swiftly digs until it can seize its target. Specialised teeth make it easy to crunch through the (often thick) armour of its insect lunch.

The ears also serve another vital function, one that's particularly important if you live in a hot environment and you're covered in fur: heat regulation. Like those of the elephant (see page 48), the fox's ears serve as radiators. Because of the fine capillary veins close to the surface of its ears, the fox can remain at a comfortable temperature by venting excess heat from its blood into the air.

Apart from its large ears, the bat-eared fox also has more teeth than other species of fox (between 46 and 50 teeth in total). In fact, these unique creatures have more teeth than most mammals. Bat-eared fox has three upper and four lower molars on each side of the mouth, while other members of the canid (or dog) family have two upper and three lower molars. These extra teeth mean that the fox is perfectly adapted to chewing creepy crawlies.

Although occasionally hunted, the bat-eared fox is one of the increasingly rare examples of an animal with a safe conservation status. It is helpful in keeping down the number of termites, which are considered pests. This means that the biggest commercial value for humans comes from keeping these foxes alive. Most threats to this fox come from the dangers its food sources might face, such as flooding from damming or the use of insecticides.

Camelus

Camel

*L*ong before humans crossed the desert, camels braved this arid landscape, surviving the immense heat thanks to their unique features that enable them to tolerate extreme heat and cold.

There are three species of camel living today and most have been domesticated, used as pack and saddle animals. The dromedary (*Camelus dromedarius*) was domesticated around 3,000 BCE and is found in dry regions in Asia and Africa, including the Sahara Desert. The domestic Bactrian camel (*Camelus bactrianus*) is found across Central Asia, and the wild Bactrian camel (*Camelus ferus*) in Mongolia and China, roaming the Gobi Desert.

With its distinctive hump, the camel is one of the most recognisable animals in the world. The dromedary has one back hump, and the domesticated and wild Bactrian camels have two humps. The camel's hump is one of the most amazing evolutionary adaptations, enabling it to go for long periods without water. But instead of storing water, as many people believe, a camel's hump is made of fat, which can be converted into energy when it doesn't have access to food or water.

If infrequent access to food and water weren't challenging enough, sand is very difficult to walk on, with each step using a lot of energy. Sinking into the sand is something we've probably all experienced when walking on the beach, but imagine how much effort it would take to walk across the desert. To help stop them from sinking into the ground, camels have wide feet and two toes that spread apart to distribute their weight more equally and therefore conserve energy.

Besides their humped backs and special feet, camels have also evolved to be able to breathe and see in hot and sandy deserts. A camel's nose works like a dehumidifier. The air a camel breathes in passes over mucous membranes that cool it, allowing the camel to better absorb moisture in the air. Camels can also close their nostrils to keep sand from getting in!

Camels have not one, not two, but three eyelids. This third eyelid, called a nictitating membrane, is transparent and can be used to wipe debris out of a camel's eye. Two of the eyelids have eyelashes — the longest eyelashes in the animal kingdom — which help protect their eyes from sand.

Sadly, there are now very few camels living in the wild. These animals are under threat from hunting and habitat loss and face competition for food with other livestock, such as llamas and alpacas. Wild dromedary camels are extinct, and the wild Bactrian camel is critically endangered.

Nautilus pompilius

Chambered Nautilus

For the nautilus, the term 'living fossil' is perhaps appropriate in a very broad sense, but the reality is there is no such thing.

The nautilus is sometimes referred to as a living fossil, as its outward appearance is little changed from that of its ancestors. But like all creatures that this term is applied to, the changes are often imperceptible. Animals such as the nautilus that on the surface appear to have stopped evolving have not been 'perfected' by evolution; rather, they have adapted so well to their environment that there has been little evolutionary demand for 'gross', or major, changes to their bodies.

During the Earth's history, there have been many extinction events that have wiped out huge portions of animal life on the planet. However, such events — whether subtle and spanning huge time frames, or sudden and drastic, such as an asteroid impact — leave pockets of life relatively untouched. An animal could be wiped out in one habitat while those in a nearby area could be relatively unharmed. These survivors will adapt if necessary and continue. Such is the possible history of the nautilus. Within its environment, it has adapted well and has weathered the events that wiped out its cousins or less-hardy neighbours.

So, with its environment remaining stable enough that 'required' evolutionary changes are only very slight, the nautilus has remained essentially unchanged for the last 500 million years. Its extinct relatives include one of the most recognisable fossils: the ammonite, examples of which can be found on almost any beach in the world.

The nautilus is a cephalopod (literally meaning 'head foot'), a member of the family of animals that includes the octopus, squid and cuttlefish. In comparison with cephalopods with short lifespans, the nautilus is extremely long-lived, possibly exceeding 20 years.

Unlike other cephalopods, the soft body of the nautilus resides within a chambered shell located behind the largest front chamber. The other compartments within its shell allow the nautilus to work almost like a submarine. The nautilus can pump water in and out of these chambers, adjusting its buoyancy and using this same pumping action to propel itself though the water. Over the thousands and million years of the species' survival, that air pocket proved such an advantage to it that it became its defining feature.

Living deep in the Indo-Pacific Ocean (700 metres below sea level), the nautilus is an opportunistic hunter and scavenger. It uses the small, parrot-like beak located in the centre of its tentacles to break into and eat its primary food source of crustaceans. When threatened, the nautilus can use two small, specially adapted tentacles to close the hood that lies on top of its head, withdrawing into its shell to protect itself from predators.

The exact population numbers of the nautilus are unknown, but they are considered to be in decline. They are caught for their shells, which are sold as souvenirs and decorative items, and also for the nacre, or mother-of-pearl, the inner layer with a pearl-like iridescence, which is harvested for its use in jewellery.

Chamaeleonidae

Chameleon

It's difficult to know where to start with chameleons. They have undergone an extraordinary number of adaptations, which relate to eyes, tongue, feet, tail and skin.

There are 202 different species of chameleon spread across Africa, Madagascar, southern Europe and southern Asia. They come in a huge range of sizes and colours, from the Malagasy giant chameleon that grows up to 70 centimetres long to the tiny *Brookesia nana*, also of Madagascar, that measures just 3 centimetres. Incidentally, this tiny chameleon could well be the smallest reptile on Earth!

Chameleons are perhaps best known for their ability to change colour. This adaptation, contrary to popular belief, isn't necessarily for camouflage. It depends on which chameleon you're referring to, as some use it for camouflage more than others, but the main reason for any colour change is social signalling. By changing colour, a chameleon can communicate how it is feeling to any other chameleons that might be nearby. It also comes in very useful for thermal regulation. Being cold-blooded reptiles, chameleons need to be able to regulate their body temperature. They can become darker to absorb light and heat, therefore raising their temperature, or they can lighten their colour to reflect light and heat, thereby cooling themselves.

Another adaptation that can't be ignored concerns their eyes, which are perhaps the strangest eyes of any animal. They are able to swivel independently, viewing two different objects simultaneously and giving the chameleon a full 360-degree area of vision. This is a huge benefit to any animal, both for hunting and for its own safety.

Chameleons live an arboreal existence, meaning that they live in and among trees. They have evolved an extremely effective set of feet to help them with this lifestyle – over time, the chameleon's toes have fused together into two sections, creating a clamp shape. As they slowly walk along, each foot can firmly wrap itself around the branch in an extremely strong grip. Many species are also aided by a prehensile tail – a tail that can be used to hold on to branches. They can use this as a fifth limb, making them even more secure.

Lastly, their tongues have to be mentioned. All chameleons are essentially insectivores, meaning that they live off insects such as crickets and beetles. Insects, especially those with wings, can be extremely hard to catch, but evolution has equipped the chameleon to be an unparalleled hunter. Its tongue is roughly two times the length of its body. In smaller species, it can be even longer. The end of the tongue is bulbous, wet and sucker-like and can be launched from its mouth at truly incredible speed. Using its unique eyes to spot a tasty-looking insect, the chameleon shoots out its tongue and, within a tiny fraction of a second, the prey is firmly grasped. Equally quickly, the chameleon will retract its tongue back into its mouth, with the tasty morsel attached.

The mechanism of the chameleon's tongue, its eyes, its skin and all its other evolutionary adaptations are extremely complex, as are the processes of natural selection that formed them. Chameleons really are a perfect example of how animals, over time, can adapt to perfectly suit their environment.

Chameleons aren't endangered and are increasingly being bred as domestic pets. However, there are some species, such as those in Madagascar, that are under threat due to the continued devastation of forests.

Sinocyclocheilus anatirostris

Duck-Billed Golden-Line Barbel

*T*here are places on this planet that are very different from the environments with which we are familiar – places where you wouldn't imagine life to exist at all.

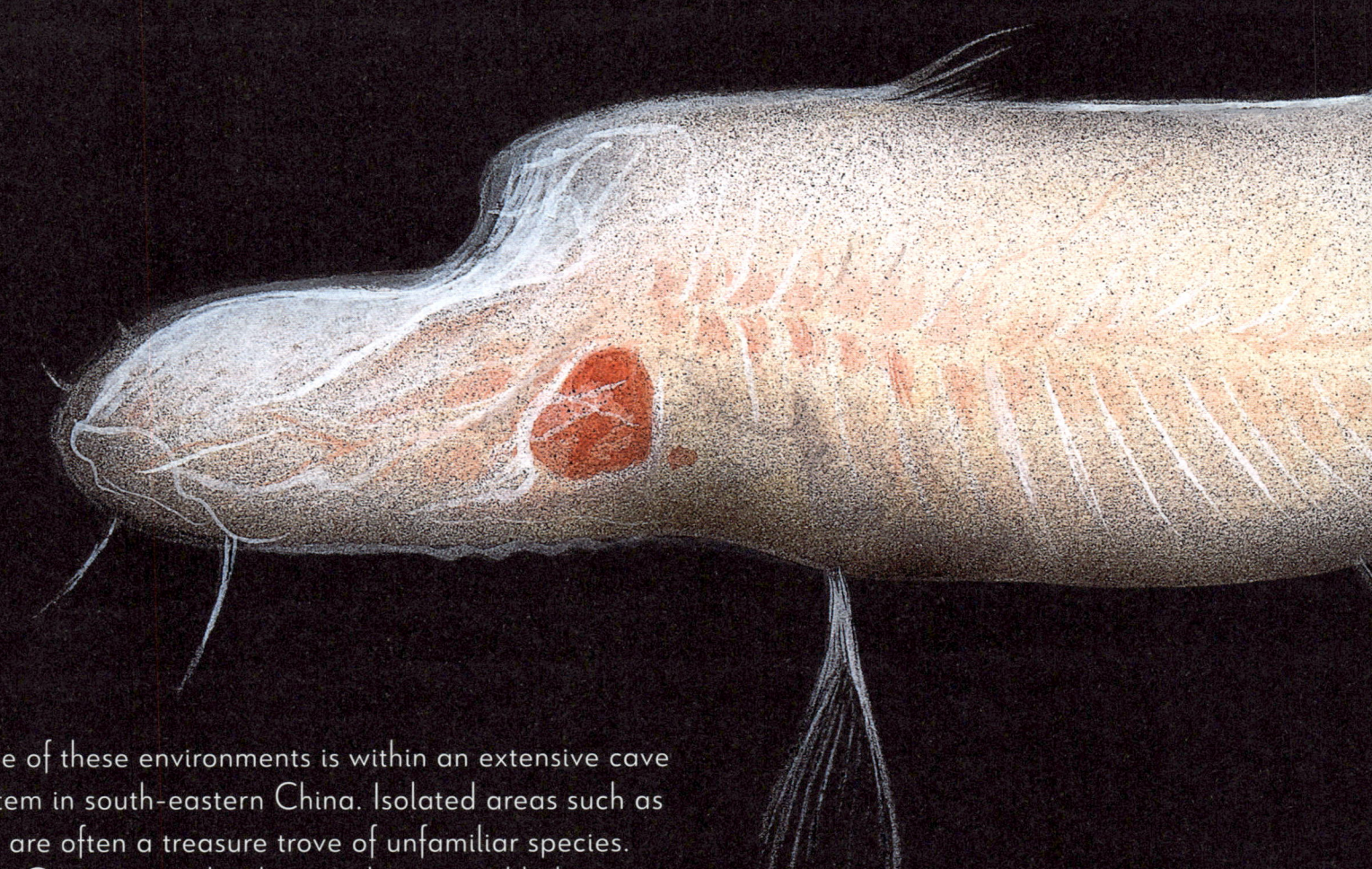

One of these environments is within an extensive cave system in south-eastern China. Isolated areas such as this are often a treasure trove of unfamiliar species.

One species that lives in this unusual habitat is a small and extraordinary fish: the duck-billed golden-line barbel. Very little is known about this fish, as its home is so inaccessible and there are so few of them left in existence. Belonging to the same family as minnows and carp, this fish used to be fairly widespread, but owing to pollution and contamination of its environment, it is now critically endangered. In just a few years, its population has declined by 80 per cent.

What makes this little fish so unique is that it is completely blind. Living in the total darkness of its cave environment has resulted in the total loss of its eyes (and its skin pigmentation). This is a perfect example of what might be considered the 'efficiency' of evolution, and how it can remove any features that are not needed. If an animal begins to adapt to a certain environment, natural selection will 'favour' the surviving animals that are able to thrive within it.

All living things require energy to function. In the case of the duck-billed golden-line barbel, the energy needed for sight has been saved over time through evolution – as they live in darkness, seeing is simply not necessary. Instead, this fish feels its way around using the barbels, or feelers, on its chin and mouth. The absence of good eyesight is an adaptation that can often be found in lightless environments such as caves and the deep sea, as we can see in the case of the naked mole rat (see page 54). Although the duck-billed golden-line barbel is a small creature (it measures up to 3 centimetres long), it is worth paying attention to it because it's in such extreme danger. Its cave home has been pristine for countless centuries, allowing this fish to adapt and thrive within it. Now, owing to human actions and pollution, it is on the brink of extinction.

Ornithorhynchus anatinus

Duck-Billed Platypus

The platypus is perhaps one of nature's most fascinating and confusing animals.

The platypus is one of five remaining species belonging to a group called monotremes. This order of animals broke away from mammals 166 million years ago. Now, all that remain are the platypus and four species of echidna (a spiny, anteater-like animal). All of these species live in Australia or on the island of New Guinea.

One of the many characteristics of the platypus, as a monotreme, is that it lays eggs. These eggs hatch and then the mother suckles its young, as would any other mammal. But, instead of secreting milk through a nipple, the platypus does so from glands on its stomach. Its young then suck the milk from the folds of its skin.

The platypus is typically found in water, where it looks for its food of freshwater shrimp, crayfish and insect larvae. It is very well adapted to this aquatic lifestyle — thick fur makes it streamlined and provides insulation, and webbed feet make it a superb swimmer. When on land, the platypus can retract the webbing on its feet, so it can easily pass through undergrowth.

The platypus is also one of the very few venomous mammals in the world. The males have spurs on their hind legs which can deliver a sting strong enough to seriously disable some animals, including humans. This adaptation is thought to be more for fending off rival males than for defence from predators.

The duck-billed platypus's bill is a soft, leathery organ packed with nerve cells, making it extremely sensitive. It also contains a series of cells called electroreceptors. The platypus can use these cells to detect its prey by picking up the tiny electric pulses produced by other animals' muscles. When it dives, the platypus closes its eyes, ears and nostrils and depends entirely on this sensitivity in its bill to hunt. Moving its bill in a rapid side-to-side motion enables the platypus to narrow down the direction of any electric impulses it is detecting. The platypus is the only mammal with the ability to use electrolocation — in fact, this behaviour is similar to that of the electric eel (see page 28).

Fortunately, platypuses are not considered to be in any particular danger thanks to conservation work. However, they are at risk from climate change and habitat destruction.

Electrophorus electricus

Electric Eel

Not that I imagine you would, or would wish to, but should you happen to meet an electric eel, I strongly recommend that you don't touch it, or even go near it. It will end badly.

First of all, these animals aren't actually eels. They're related to a family of fish called knife fish. They just happen to be very eel-like and therefore have become known by this name. Living within the waters of South America, the electric eel is extremely large, thick and muscular and can grow to 2 metres in length. Of the three species of electric eel, the *Electrophorus electricus* was the first to be discovered.

Despite living underwater, electric eels gain 80 per cent of their oxygen by coming to the surface and breathing in air. Although they do have gills, this is not their primary source of oxygen intake, which means that they are able to survive comfortably in water that has a very low concentration of oxygen.

Obviously, the reason this fish is included here is that it produces truly huge amounts of electricity; and not only that — it is also able to produce both high and low voltages. It has evolved this ability as a means to locate, stun and even kill its prey, as well as to defend itself, which it can do very effectively.

The eel's body contains three different organs that produce this power. What is fascinating is that they literally work like a battery. These organs are made of cells called electrocytes, which are modified nerve or muscle cells. These are lined up and stacked, just like the cells in a battery, increasing and storing the electrical power.

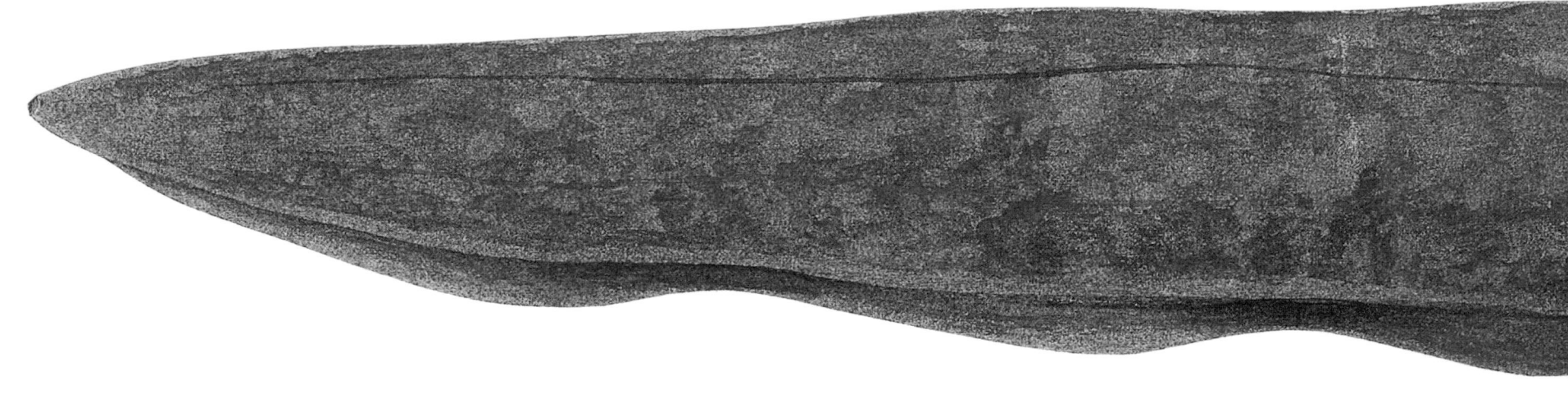

The way eels use this power is remarkable. Not only can they produce a sudden discharge of power to stun prey or deter a predator, but also they can use this electricity essentially to 'remote control' their prey. By varying the rate and power of their electrical discharges, they can cause a reaction in the nerves of other fish. This in turn can cause the other fish to twitch or move, revealing their location or even causing them to swim closer to the eel.

Although electric eels are considered aggressive animals, they are not. They primarily use their electric organ for electrolocation as well as identification of objects underwater. This is especially important because of their poor eyesight. Eels are nocturnal animals that live in dark waters, so they must rely on electricity to help them 'see' their surroundings.

The electric eel lives in rivers as well as in muddy streams and stagnant water, and it is not considered to be particularly under threat. However, increasing levels of pollution in these waters as well as climate change are inevitably going to have an impact. While we may fear them, we must also marvel at the amazing abilities eels have evolved in order to survive.

Talpa europaea

European Mole

These small, unassuming animals are usually considered pests. Bearing in mind that they live underground, only really emerge at night and live on a diet of earthworms, insects and other small creatures, you might wonder why moles have this reputation.

Moles are killed in the thousands every day because they produce mounds on people's lawns. Sadly, this is just one of the many examples of how humans can devastate populations of animals simply because they cause a supposed inconvenience.

The mole's most recognisable attribute may be its enormous front paws. If you're going to spend your life underground, it's extremely useful to have a spade. Evolution has provided the mole with two of these in the form of modified front paws that are essentially two huge scoops (similar to the pink fairy armadillo; see page 66). Moles burrow almost constantly, creating huge networks of tunnels in order to find food. In fact, a mole can dig up to 20 metres of tunnel a day!

The mole has evolved various features to suit its environment and lifestyle. One such feature is poisonous saliva, which it uses to paralyse earthworms and insects, making them easier to consume.

Since they are nocturnal and live mostly underground, moles don't really have a great deal of use for good vision. As a result, their eyes are extremely small and hidden under their fur. Although they are almost blind, moles are expert hunters underground, using their acute senses of touch and smell when burrowing to detect other moles and prey. Unlike their cousin the shrew, which primarily relies on its whiskers to detect food through touch, moles use the sensitive skin on the tip of their nose to explore their tunnels and search for prey.

Moles mainly use their nose to smell, however — their sense of smell having evolved to be far more advanced than their other senses. In fact, studies have shown that moles smell in stereo, which means that they use their nostrils independently to sniff out scents and paint a picture of their dark surroundings, enabling them to locate food, navigate tunnels and sense other moles.

Because they spend their life underground, moles are less often captured and consumed by predators, such as birds of prey and foxes. However, although moles are not considered endangered, they are still targeted for hunting. No longer widely hunted for their pelts, moles are commonly killed as pests, particularly on livestock farms, where horses and cattle can break their legs by accidentally stepping into molehills.

Despite these little animals being considered a nuisance by many, moles benefit the ground in a similar way to worms, aerating the soil by churning it up. They also eat a large number of insect species that are considered pests by gardeners. With any luck, people will learn to appreciate that a molehill in the garden is something to be treasured — if you find one, it means that these animals have chosen your garden to call home.

Anoplogaster cornuta

Fangtooth Fish

Despite its fearsome appearance, the fangtooth is a fairly small creature and is completely harmless to humans.

The fangtooth lives deeper in the ocean than almost any other fish we know of. Although typically found between 300 and 2,000 metres down, they have been found as deep as 5,000 metres below the surface. Of the two known species, the largest grows only to about 15 centimetres, and the other is thought to be half this size.

Living in the pitch-black, frigid and inhospitable waters of the world's deep oceans often leads to an incredible range of adaptations in the inhabitants there, such as the anglerfish and daggertooth, to name just two. Many of these adaptations can be seen in the features of the fangtooth.

Because of the almost total absence of light, the fangtooth has very poor eyesight, but unlike some animals that live entirely in total darkness, such as the duck-billed golden-line barbel (see page 24), it has retained its eyes as they are still of some use.

During the night, the fangtooth drifts gently from the deep towards the surface, taking advantage of the thin moonlight filtering through the water. There its eyes, however poor, may be an advantage.

It's understood that the fangtooth hunts predominantly using a method called chemoreception, which is a very impressive way of saying 'bumping into things'. Other animals that use this method of hunting also share many of the fangtooth's traits, the most notable of which are its teeth. In proportion to its size, the fangtooth has the largest teeth of any fish in the ocean.

As you can imagine, if you live almost entirely in the dark and you can only find food by bumping into it, it's vital to your survival that your jaws are able to grab and hold on to prey instantly. Therefore, over the many millions of years of its evolution, the fangtooth has developed exactly the right tools for its survival.

To accommodate these huge, dagger-like teeth, nature has also provided it with two grooves on either side of its skull that allow the teeth to slide along, so that the fish can close its mouth at least partially.

As are all fish, the fangtooth is equipped with a sense organ called the lateral line. This feature runs along the side of a fish and allows it to detect pressure, movement and vibrations. When you live in almost total darkness, this organ is of incredible importance, and in the fangtooth it's understood to be particularly well developed.

Cryptoprocta ferox

Fossa

At first glance, you might wonder what exactly a fossa is. It's a little like a dog, kind of like a cat and somewhat similar to a mongoose – and that is precisely why it's so interesting.

The fossa lives predominantly within the (rapidly disappearing) forests of Madagascar, sleeping, hunting and mating among the branches. As the top predator on the island of Madagascar, it has evolved a number of traits, some of which are found across various different species.

Superficially, the fossa resembles a small version of a puma. Its slender body, covered in short, dense fur, is roughly 80 centimetres long. Its muscular legs are equipped with heavy claws that are semi-retractable, meaning that it can withdraw or extend them (like a domestic cat). It also possesses unusually flexible ankle joints. This might not sound particularly noteworthy, but this adaptation allows the fossa to climb up trees rapidly. Just as importantly, its claws, strong muscles and specialised ankles allow it to run *down* trees just as well. Together, these features make it an extremely efficient hunter of equally well-adapted prey, such as lemurs, rodents, birds and lizards.

The fossa is also equipped with a long tail that helps it to balance while climbing and hunting among the tree branches, where it spends most of its time. Its extra-long tail, which can grow up to the 70 centimetres, almost the same length as the fossa's body, helps the animal balance and jump from branch to branch.

When female fossas reach around one year of age, they start exhibiting a truly remarkable change to their body – they become temporarily masculine. Although they don't become male, they develop masculine traits. Their genitals change shape, and they begin to emit an orange substance from their glands that changes the colour of their fur so they resemble a male. The advantage of this change has long been a mystery, but it has been suggested that female fossas have evolved in this way to help prevent them from both being sexually harassed by males and experiencing aggression from other females while they are still young. As the animal ages, these changes stop, and it reverts to the characteristic female form.

The fossa is currently considered under threat and possibly endangered owing to the continual destruction of its forest habitat. Sadly, Madagascar has lost 80 to 90 per cent of its original forest cover, which means that this beautiful animal is at risk of losing its home entirely.

Myrmecophaga tridactyla

Giant Anteater

There is no getting away from the fact that the giant anteater (or ant bear, as it's sometimes known) is an extremely strange-looking animal.

This anteater can be found throughout Central and South America, in habitats ranging from open grasslands to shady forests — wherever it can find the ants and termites that form the main part of its diet. In fact, this diet has been central to the evolution of the giant anteater. Ants and termites, which live in huge colonies of many millions, are an abundant food source full of protein and nutrients. However, these colonies often make their homes in immensely strong structures of dried mud and earth — they also often live underground. This makes it extremely hard for predators to reach them. As a result, any animal that has adapted to be able to track them down has a huge advantage and is more likely to survive. The giant anteater has done this in no uncertain terms.

Perhaps the most obvious feature the giant anteater has developed is its snout. This anteater possesses a very acute sense of smell that allows it to locate termite mounds and anthills — it can even use its nose to identify the type of insect they might contain. The giant anteater's snout is long and narrow, forming a curved tube. It does have a jaw, but it is completely toothless. All of this allows the anteater to squeeze its snout into a relatively narrow space — important when your food is encased in something not unlike concrete. The less energy you need to expend to reach your food the better, so if your snout can adapt to fit into a hole you didn't have to create, then a great deal of time and energy is saved.

The anteater feeds with its tongue. This tongue is incredibly long, extending almost 60 centimetres beyond the end of its snout, and it is covered in tiny barbs and a thick, sticky saliva. With its ability to flick this tongue in and out up to 150 times a minute, the anteater is able to consume around 20,000–30,000 insects every day.

As it doesn't have teeth, the anteater needs some way to grind up the insects to release the nutrients they contain. It has evolved a stomach containing strong muscles and powerful acids that crush up and dissolve the food, replacing the need for teeth in its jaw.

If all this wasn't remarkable enough, the giant anteater also possesses three large powerful claws on each front foot. It uses these mainly to break through the thick walls of termite mounds or to dig into the earth to find ants. However, they are also a fearsome weapon, which the anteater can use if it needs to defend itself against predators such as big cats.

Some anteater species have a prehensile tail (one that is capable of grasping onto objects by itself.) The anteater can use this as a fifth limb, for gripping branches or providing support if it needs to stand on its hind legs to feed. Also, the anteater can fold this magnificent bushy tail around itself, providing a great deal of warmth – very useful, as its habitats can be extremely cold, especially at night and in colder seasons.

This truly fascinating creature is currently listed as vulnerable owing to wildfires and the destruction of its habitat for logging and housing. It is also affected by poaching and people hunting it for its fur and meat. Upsettingly, it is also sometimes hunted for sport. It is to be hoped that with time, and as people become more aware of how extraordinary it is, this anteater's future can be assured.

Numida meleagris

Helmeted Guinea Fowl

The helmeted guinea fowl is visually extraordinary. Instantly recognisable, it is unfortunately often cited as one of the world's 'ugliest' birds, but the reality is that it is highly intelligent and social.

The most notable adaptation this beautiful bird possesses is its 'helmet', also known as a casque – the colourful boney bulge at the top of its head. This may serve as a radiator or 'thermal window' in much the same way as an elephant's ears do (see page 48), allowing the guinea fowl to cool down easily. Evolving as this bird has in a hot, arid environment (it is native to the Sahara Desert), this is a much-needed adaptation, and the horn-shaped helmet is perfect for this purpose.

The helmeted guinea fowl's plumage is also perfect for its environment. Small feathers with regular polka dots give the bird superb camouflage – in sunlight and darkness, open ground or thick bush; each feather on its body has dozens speckles.

A guinea fowl's head is unfeathered, like a vulture's, and its warty facial skin is coloured red, white or blue, depending on the bird's geographic location. It also has two wattles, one on each side of its beak, which are usually red or blue.

The helmeted guinea fowl is approximately the size of a chicken, with a small head and a large, round body. It is often thought to be flightless, yet it is perfectly capable of doing so. However, the helmeted guinea fowl prefers to walk, and only really uses its wings to avoid danger or to fly up to its treetop roosts in the evening.

These characterful birds are hugely beneficial to the ecosystem, as they reduce the number of ticks and other insects and spread seeds that allow plant life to flourish. Despite being native to the Sahara Desert in Africa, helmeted guinea fowl can now be found all over the world. They are not threatened globally and are widely domesticated, popular for their meat and eggs.

On farms, they also make great 'watchdogs' as they give an extremely alarming cry when disturbed.

Hippopotamus amphibius

Hippopotamus

Mighty and majestic, the hippopotamus can usually be spotted in the wild submerged in water, concealing its immense size and formidable presence.

After elephants and rhinos, the hippopotamus (or hippo, as it's more commonly known) is the third-largest land animal on Earth. It can easily grow to more than 3 metres long and weigh up to around 3,200 kilograms. The hippo is semi-aquatic, meaning it lives a large part of its life in water – specifically, the river waters of its home in sub-Saharan Africa. In fact, other than when it's eating, it pretty much lives its entire life in the water. During the day, the hot sun can damage the hippo's delicate skin, but in the evening and at night, when it's cooler, it will leave the water to feed. Hippos often travel several miles at a time in search of the tasty grasses that constitute their diet.

As beautiful and extraordinary as these animals are, do not be tempted to approach one. The hippo is in fact one of the most dangerous animals in the world. Aggressive and alarmingly unpredictable, hippos attack humans much more often than do ferocious-looking predators such as lions. They have a bite that is twice as strong as a lion's and only a little less strong than the saltwater crocodile's, which has the most powerful bite of any animal (see page 74).

The hippo's aquatic lifestyle means that it has evolved certain characteristics. Its legs are relatively short and squat because the water supports most of its body weight. The legs might be short, but don't be mistaken — a hippo can run at great speed. In some cases, they have reached 50 kilometres per hour over short distances. They can also make quite surprising leaps up river banks, so if you see one while strolling along, it's best to keep your distance!

Despite spending so much time in the water and even having webbed feet, hippos can't actually swim in the strictest sense. Instead, they will push themselves off the riverbed and glide from spot to spot. They stick to fairly shallow water, and, even when they're entirely submerged, they sometimes just walk or pull themselves along the bottom.

Hippos are also known for their incredible tusks. As these animals are herbivores that feed on plants in the water and on the riverbanks, these teeth are not used for eating. In fact, hippos have developed tough, almost horny, lips for pulling at tough grasses and weeds. The tusks have evolved purely for fighting. Male hippos are aggressively territorial and will often fight each other both for the possession of a short stretch of river and for the dominance of groups of females. These tusks can inflict enormous damage to other animals, so a fight is not taken lightly.

The hippopotamus has become a vulnerable species because of poaching — sometimes for its meat but more often for its tusks. Like the elephant's tusks (see page 48), these are made of ivory and are therefore considered extremely valuable. So, unfortunately, the hippo is targeted by people who wish to profit from them. Hippos are also sometimes killed as they are considered a danger to humans. Though this may well be true, it is only because of the behaviour of us human beings as we encroach on the hippo's habitat in the search for territory of our own.

Bucerotidae

Hornbill

The hornbill is an intriguing type of bird because of its appearance and its behaviours.

The three hornbills shown here – (from top to bottom) the Abyssinian ground hornbill, the black-and-white-casqued hornbill and the Malabar pied hornbill – are examples of the 62 species of hornbill currently recognised. Hornbills are widely spread, inhabiting various environments in subtropical Africa and Asia.

These sensational birds are particularly distinctive because of their enormous beaks, which come in a huge variety of shapes, sizes and colours and they have a horny bump on the top of their head, known as a casque. These casques are very small and seem to be vestigial, meaning they essentially no longer serve their original purpose. This is an example of evolution removing or reducing the size of a feature that is no longer important to that animal's survival. It is thought that the casque may function as a means for male hornbills to attract females. For many hornbills, the casque also acts as a resonator, which means that the sounds they produce are amplified as they move through the hollow casque.

The hornbill's beak serves a range of purposes, from preening and feeding to fighting and constructing nests. Some hornbills even use their beaks and, in some species, thick, bone-filled casques for aerial jousts where they head-butt a rival in mid-air. Over time, several of the hornbill's neck bones (vertebrae) have fused together. Along with thick neck muscles, these bones support the weight of the beak.

Hornbills have evolved a unique nesting habit that is a great aid to the survival of their offspring in their early years. When a pair of birds have mated, they locate a hollow in a tree for the female to nest inside. The male hornbill will then close up the hole, using mud and even its own droppings, leaving a small opening for it to pass food to its mate. Inside this safe home, the female will lay up to six eggs which, when hatched, are safe from predators. When the young birds have fully matured, the female hornbill breaks open the nest wall and emerges into the daylight.

Despite this bird's amazing adaptations, the population of 51 out of 62 species of hornbill is decreasing. In the wild, hornbills are victims of poaching, as their casques are reminiscent of ivory and easy to carve, and therefore can be used to make jewellery.

Limulus polyphemus

Horseshoe Crab

The first thing you need to know about the horseshoe crab is that it isn't actually a crab at all. In fact, it's more closely related to arachnids such as spiders and ticks.

Like the chambered nautilus (see page 20), these animals are often called living fossils – a phrase used to describe an animal whose appearance has barely changed over millions of years. However, this is a slightly misleading phrase, and science frequently shows this to be untrue. An animal will continue evolving as its environment changes. These changes might seem superficial or be so small that they are almost unnoticeable. It might not be initially evident that the evolutionary process is happening, but it is.

Although the horseshoe crab's shell acts as heavy armour, it is also extremely sensitive to light. These crabs are equipped with ten eyes – a pair of compound eyes on the front of the main shell (a compound eye, as you will find on insects and crustaceans, is an eye that's made up of a great number of tiny but essentially separate eyes, each creating their own image), simpler eyes on the main body, and two additional simple eyes on each side underneath the mouth. It also possesses groups of more rudimentary photoreceptors along its tail (also called a telson).

The horseshoe crab is extremely important both within the ecosystem of its habitat and as an ecosystem in itself. A broad range of other species attach themselves to the crab – worms, molluscs, barnacles and seaweeds all utilise the crab for their own survival. They're also important as a food source. Younger crabs provide valuable nourishment for animals such as sea turtles, sharks and alligators, and their eggs are a vital food for shorebirds as they refuel on their migrations.

There are four species of horseshoe crab – the Atlantic horseshoe crab pictured here lives along the coasts of North America and the others can be found in South East Asia. The Atlantic horseshoe crab is listed as vulnerable, although the status of populations varies somewhat between different areas. The primary reason for their vulnerability is overharvesting. They are used as bait by fisheries and also, significantly, for medical purposes. Their blood is copper-based (unlike our own, which is iron-based) and contains a substance that will coagulate, or thicken, when exposed to toxins. As a result, their blood can be used as an effective test to check whether medical equipment, vaccines and injectable drugs are sterile before they are used. Although the extraction of the crab's blood is non-lethal in theory, the mortality rate of the crabs who are 'bled' is often as high as 30 per cent. Owing to the conservation status of these animals, there are some efforts under way to find other means of carrying out this test that don't require the crab at all.

Threskiornithidae

Ibis

Due to its elegant and regal appearance, it's no surprise that in ancient Egypt the sacred ibis was worshipped.

Thoth — the Egyptian god of the Moon, magic, wisdom, knowledge and writing — had the head of an ibis, and so these majestic birds (whose curved beaks resemble a crescent moon) were believed to be the living incarnation of Thoth on Earth. Ibises are depicted frequently in Egyptian hieroglyphics, and the birds were often mummified and then buried with pharaohs.

A species of wading bird, the ibis is commonly found in wetlands, but they also live in forested and grassland areas. These birds prefer tropical and subtropical habitats and can be found around the world in all warm regions except on the South Pacific islands. Ibises are social creatures and travel, feed and breed together in large flocks, which offer protection from predators. But it's an adaptation they developed that makes them particularly unique.

All ibises have long, downcurved bills as well as long legs and toes that make it easy for them to wade through mud and low water to find food. Their bills are specially designed with their nostrils located at the base of the bill instead of the tip so that they can breathe while locating their meal in the water.

If that didn't give them enough of an advantage, their bills also have sensitive feelers, which help them identify food that they find while they are wading through long grass. Most interestingly, baby ibises' bills are straight at birth and start curving downwards around 14 days after birth as they continue to grow.

Their appearance varies among different species, and ibis feathers can come in a variety of colours, including white, black, brown, grey or even pink. Most ibises have bald heads, which are usually black or white, but the skin turns bright red during the breeding season. Ibises' colouring is also determined by their dietary habits and habitat. As is the case with flamingos, the scarlet ibis's bright pink colouring is a result of the algae and crustaceans it consumes.

There are currently 28 species, with several of them classified as endangered. The destruction of the ibises' habitat is largely to blame, as commercial logging and the draining of wetlands have destroyed their homes. Six species have already become extinct, including two that were flightless birds – *Apteribis* of the Hawaiian islands and *Xenicibis* of Jamaica, the latter of which had uniquely shaped club-like wings. Without the protection of this creature's habitat, we could see even more species of ibis disappear forever.

Elephas maximus indicus

Indian Elephant

Familiar as we all are with elephants, it's important that we don't take them for granted. When looking at an animal such as this with fresh eyes, you begin to see the truly extraordinary features that the processes of evolution have generated.

Elephants are the largest land mammals on Earth, having evolved to become enormous in size as a way to help them survive in niche environments where the vegetation has little nutritional value (see also American bison, page 10). The *Loxodonta africana* or African savanna elephant is the largest species of elephant, growing to be a shoulder height of 4 metres and weighing more than 5,000 kilograms.

The most recognisable feature of the elephant is, of course, its trunk. Containing 150,000 muscles, the trunk has a huge array of uses. The elephant uses it to breathe, smell, grasp, reach, gesture and fight. It is also vital for feeding, drinking and bathing. Yet despite its enormous strength, the trunk also possesses incredible sensitivity. It is powerful enough to enable an elephant to rip a tree from the ground, but it can also crack a nut without damaging the seed within.

The elephant's ears have also adapted to suit its environment, growing to act as radiators. Close to the skin's surface, the ears are densely packed with blood-carrying capillaries. When the ears are flapped, they allow the elephant to shed excess heat efficiently into the air. Indian elephants, like the one shown here, live in much cooler environments than their African relatives — as a result, they have developed smaller ears.

As well as its obvious physical attributes, it's important to note the elephant's mental characteristics. They are incredibly social and live in family groups, and studies of elephants have indicated clear signs that they are highly emotionally sensitive. Importantly, like humans, they can also recognise themselves in a mirror (rather than assuming their reflection is another elephant), which is an indicator of self-awareness.

Sadly, throughout history and to the present day, elephants have been murdered for the ivory of their tusks, or because their home overlaps with human territory. Thankfully, some wonderful organisations are doing amazing work for elephants, making numerous and constant efforts to try to protect these breathtaking creatures.

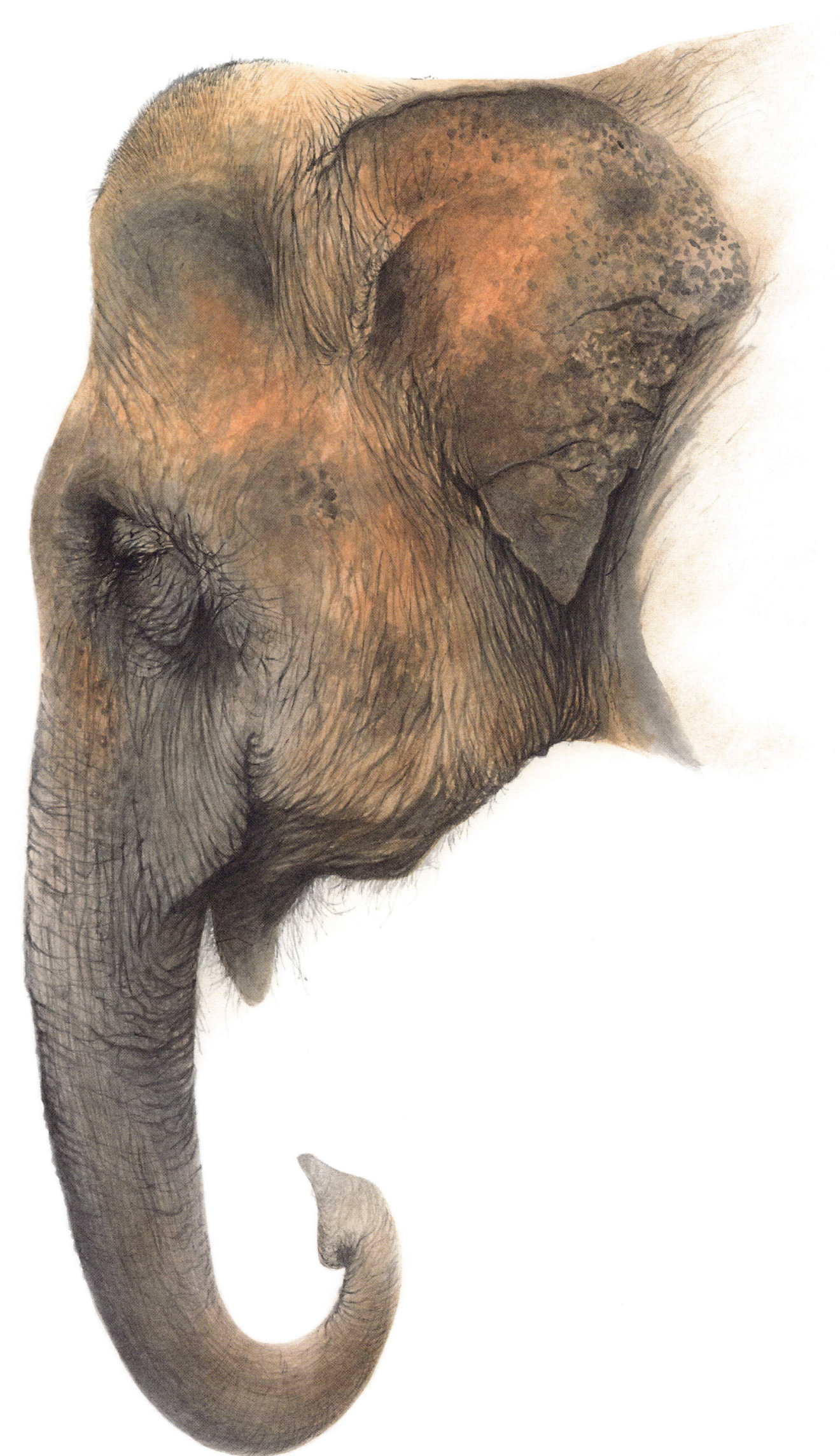

Manis crassicaudata

Indian Pangolin

Pangolins are truly gorgeous animals. Its armoured scales look like something from prehistoric times.

There are eight species of pangolin, which all look relatively similar. Their defining trait is, of course, their scaly hide. The pangolin is covered in thick, broad, interlocking plates of keratin, the same substance that rhinoceros horns, tortoise shells and our own fingernails are made of. It is an extremely tough substance that is perfect for protecting the animal. The interlocking structure of the plates also makes the pangolin very flexible — when threatened by a predator such as a tiger, the pangolin is able to curl itself into a ball, protecting its soft belly.

The pangolin is sometimes called the scaly anteater and shares the same diet as the giant anteater (see page 36). Both eat insects, such as termites and ants — animals with nests that are often incredibly difficult to enter. To reach their food, the pangolin and anteater have developed extremely strong and long claws that allow them to excavate these nests in search of nutritious ants. However, despite their shared diet and other shared adaptations, these two animals aren't actually related. This is a prime example of convergent evolution, where multiple animals have adapted in very similar ways to accomplish certain tasks.

The pangolin also possesses a long, sticky tongue that can whisk in and out of the ants' tunnels, sucking up the small insects into its mouth. This tongue really is an astonishing piece of equipment. It's so extremely adapted to the pangolin's diet that it is actually longer than the animal's head and body combined. Also, unlike other animals' tongues, the pangolin's isn't connected to its mouth or throat. It is connected at the base of the ribcage instead. When it's not being used to vacuum up tasty ants, the pangolin's tongue rests in its chest. The pangolin doesn't possess teeth as it has no requirement to chew its food. However, it does have extremely powerful stomach muscles that help churn and break up its food, another example in which it is similar to an anteater.

As it is a nocturnal animal that forages for its food at night, the pangolin isn't particularly well studied, which is a shame as it's such a fascinating creature. Sadly, pangolins are one of the most trafficked and threatened animals in the world. They are hunted and sold for meat, and some cultures still insist on the idea that their scales (which are completely dead and inert) have medicinal properties – they do not. They are so under threat, in fact, that Asian pangolin populations have shrunk by a staggering 80 per cent in the last 10 years alone. This is a tragedy, and it ultimately means that we are in great danger of losing these creatures forever.

Tapirus indicus

Malayan Tapir

The Malayan tapir, found in the forests and rivers of Sumatra and Malaysia, Thailand and Burma, might look as if it could be related to elephants, with its long mobile nose, but it is in fact much more closely related to horses and rhinos.

The tapir's most recognisable feature is its nose, or more accurately, its proboscis. Although visibly and structurally different from an elephant's trunk, the tapir's nose is used for very similar tasks. Some scientists consider a tapir's snout and an elephant's trunk to be an example of parallel evolution, meaning two unrelated species developing a similar adaptation in a particular environment.

A tapir's snout, which is a fusion of its nose and upper lip, is extremely flexible and able to move in all directions. This allows the tapir to reach and hold branches and leaves that would otherwise be inaccessible. When swimming underwater, the snout can be used as a snorkel. The snout provides a very fine sense of smell, which is a great advantage as the tapir generally has very poor eyesight.

The tapir's otherwise usually brown eyes frequently show a blue cloudiness, essentially a 'fogging' effect, that comes from too much exposure to bright sunlight. It has been suggested that this deficit in one of the senses has become one of the driving forces for the development of such a sensitive sense of smell and hearing – so that a kind of evolutionary compensation has taken place.

An adaptation the tapir possesses that is seen only in its young is its camouflaged coat. The adults have few natural predators, but the young animals are in danger from big cats, such as tigers and jaguars, so evolution has provided them with a remarkable disguise of a brown coat with white stripes and spots, which allows them to become almost invisible in the dappled sunlight of the forest floor.

Although quite a large animal, often weighing in at 320 kilograms, the tapir can be extremely quick and agile on its feet and is just at home in water as it is on land. In a similar way to hippos, tapirs will swim out into the water, sink to the bottom and often walk along the riverbed, feeding on the succulent water plants growing there. This behaviour also benefits them, as it allows river fish to clean them, picking parasites and dead skin from their bodies.

A unique feature compared to other mammals, tapirs have four toes on their front feet and three toes on their hind feet. The four toes at the front allow the tapir to walk with ease in the soft, boggy mud of the forest floor or the bed of a river as it seeks out the leaves, fruits, nuts and water plants that form its diet. As with many species, the tapir plays an essential role in its environment by spreading the seeds of its favourite foods throughout the forest.

Sadly, all tapir species are in danger of extinction. Numbers are dwindling fast as a result of the destruction of their habitat through deforestation; the damming of rivers, which causes their environment to flood; and illegal hunting.

Heterocephalus glaber

Naked Mole Rat

The naked mole rat is the longest-living rodent species in the world!

Living in burrows within the hot earth of eastern Africa, the naked mole rat has no real need for fur; as a result, natural selection has removed hair from most of its body. However, the naked mole rat is not completely hairless. It has sensory whiskers on its face and tail, and they have hairs between its toes which allow its feet to function as brooms to sweep soil away.

These underground creatures have also evolved to be the only mammals that are almost entirely ectothermic, meaning cold-blooded. Unlike most other mammals, naked mole rats cannot maintain a steady body temperature. However, on the rare occasion that they do become chilled, they huddle together for warmth, which is easy when they live in colonies of up to 300 individuals.

This isn't the only way evolution has touched the mole rat in an extraordinary way. As well as lacking fur and warm blood, they have eyes that are almost sightless (their tunnels are entirely dark), they can survive in incredibly low-oxygen environments (to the point where any other creature would suffocate), and they do not feel pain in their skin. They are resistant to cancer and are incredibly long-lived; some individuals have lived for more than 30 years, which is a very long life for a rat.

Unlike other burrowing creatures such as the European mole (see page 30), the naked mole rat uses its teeth, rather than its feet, to dig its tunnels. As a result, almost a quarter of its entire musculature has developed in its head for the purposes of opening and closing

its jaws. Also, it is able to move its front teeth independently of one another, almost like chopsticks, which aids in excavating its tunnels.

As if all this wasn't enough, these astonishing creatures are extremely social and have a system that is more similar to that of termites or bees. There is a queen and a breeding male, with the rest of the colony being sterile workers with designated roles such as finding food, maintaining the tunnels and looking after the young.

So, this small creature, which is so often considered 'ugly', is a beautiful example of what, over time, nature does to aid the survival of a species. Although the naked mole rat has no immediate threats, it is sadly considered a pest that damages and destroys crops, and is therefore often hunted by humans. If we are to learn more about them, these unique creatures need our protection and respect.

Monodon monoceros

Narwhal

Unicorns are real, and they are far more fascinating and beautiful than any horse-like creature you might find in a fairy tale. Instead, they are a type of whale.

In many ways the narwhal is a pretty 'standard' whale. It is of medium length, around 5.5 metres long, and lives in the chilly waters of the Arctic. One standout adaptation of the narwhal is that it doesn't possess a dorsal fin (a pointed fin on the top of an animal's back, like you might see on a dolphin). This is likely to do with the narwhal's environment. It has been suggested that this fin has been 'lost' so that the narwhal can swim under large sheets of ice more easily as well as avoid losing body heat.

Its single most remarkable trait, though, is its tusk. This tusk, or horn, is in fact an adapted canine tooth. It grows from the upper left-hand side of the jaw, passes through the animal's upper lip and continues in a spiral shape, eventually growing up to around 3 metres long. There are occasions when a narwhal will grow two tusks, which appear as a forked 'V' shape, but this is exceedingly rare.

It's not certain exactly why the narwhal has developed this tusk, but if you want to know what uses it has for the creature, 'a lot' would be the short answer.

Detailed analysis shows that it's an incredibly sensitive organ. It's packed with millions of nerve cells and is also covered with tiny holes, which allow seawater to enter. This makes the narwhal very sensitive to its environment, allowing it to gauge the temperature and salt content of the water it is swimming in. It might also be helpful for locating food.

It was once thought that this tusk was also used for display and for narwhals to show aggression towards each other. However, these animals have been witnessed rubbing their tusks together. It has since been suggested that this characteristic gesture is a way for narwhals to communicate.

The tusk undoubtedly has many other uses for the narwhal, including hunting and poking holes in the ice. However, whatever purpose the tusk originally evolved for, it isn't crucial to the narwhal's survival, as the females generally don't possess one. The broad view of the reasoning behind this is that the tusk is a secondary sexual characteristic, similar to the antlers of a stag (see page 76).

Unfortunately, the conservation status of these beautiful animals is of special concern. They are being threatened by climate change and the reduction of sea ice, oil and gas production and chemical and noise pollution.

Babyrousa celebensis

North Sulawesi Babirusa

The babirusa, or deer pig, has an undeniably unusual appearance. It has thick, virtually hairless skin, a heavy snout and long curved tusks – all making it unmistakable!

In some ways, though, the babirusa is a bit of a mystery. You might assume that those extraordinary and unique tusks are for defence or foraging, or perhaps have a similar purpose to those of other tusked pigs such as the wild boar (see page 106). In fact, their purpose is not understood.

Only the male babirusa possess the longer, upper tusks, and they will keep growing continuously throughout its life. These tusks are actually the upper canine teeth of the animal, growing from the upper jaw. Initially, they grow downwards, but as they develop, they begin to curve, spiralling backwards, piercing the skin. It has been reported that, in rare cases, these tusks have actually grown long enough to curve back into the top of the animal's head. We know that these tusks aren't used for defence – even though they look fearsome, they are in fact relatively fragile. Our best guess is that these tusks are for display. The fact that only males have these features also makes it seem likely that they are there to impress females.

Their tusks aren't the only mystery that these animals hold – babirusa also engage in a puzzling behaviour called 'ploughing'. Again, it is only the males of the species that do this. They kneel and push their heads through soft ground (hard ground would damage their tusks) to create furrows. As they do this, they grunt and growl and produce large amounts of foamy saliva. It is thought that this might be a kind of power show, or a way for the babirusas to mark their scent.

The babirusa is omnivorous, which means that it is capable of eating pretty much anything. Its ability to consume many different kinds of foods is a huge evolutionary advantage, especially when food is scarce. The babirusa will happily eat leaves, insects, small mammals and – gruesomely – even smaller babirusas.

This remarkable animal is now considered endangered in its home, the island of Sulawesi in Indonesia. Although hunting babirusa is illegal, this still happens frequently. Almost inevitably, it also suffers from habitat loss owing to illegal logging. This in turn makes the usually well-camouflaged babirusa much more exposed to poachers, putting this fascinating animal at grave risk.

Dipus sagitta

Northern Three-Toed Jerboa

With its long tail and hind legs, the jerboa has evolved to be able to spring away quickly from predators, hopping around like a rabbit.

There are 33 different species of jerboa, a small rodent that can be found across a vast area encompassing Iran, Russia, China, Mongolia, Africa, India and Pakistan (and one extremely rare marsupial species in Australia). All are adapted slightly differently depending on the conditions of their own particular habitat – some have evolved to survive extreme cold, and others extreme heat. Pictured here is the northern three-toed jerboa.

Jerboas live in barren desert environments where food and water are often scarce. They are also very vulnerable to many predators. All of this has led jerboas to evolve a variety of means to further their survival. There are, of course, differences between the species – some have larger ears, for example. But what is common to all jerboas is their means of moving about.

Like kangaroos, jerboas have evolved powerful hind legs that enable them to move around by hopping. Despite its small size, its length measuring up to 10–15 centimetres, the jerboa has legs so powerful that it can bound an incredible 3 metres in one jump! These creatures have also evolved the tactic of moving quickly in a zigzag motion, and can reach speeds of up to 24 kilometres per hour. This all helps them to avoid predators and escape in an emergency.

When they're not zooming around the desert, jerboas busy themselves with finding their own food, predominantly plants, grubs and small insects. As their environment is typically very dry, they receive all their water from their food.

As we have seen with animals such as the bat-eared fox (see page 16), the jerboa has evolved superb hearing; it also has extremely good vision. Being nocturnal, they have to be on high alert at all times. The specialised senses the jerboa has evolved also come in handy when locating food in the dark.

    The jerboa's habitat is prone to huge temperature swings – often baking hot during the day and literally freezing at night. To avoid these extreme temperatures, the jerboas live in burrows, ideally in fairly high terrain so as to avoid the flooding that occurs during the rainy seasons. What is ingenious, though, is that these burrows have a secret. Jerboas are often hunted by snakes that can enter their burrows, or foxes that can dig them out. So the jerboa creates an emergency exit from its home. Should anything unwelcome decide to poke its head inside, the jerboa can immediately remove itself and live to see another day.

    Despite their environments being under threat from climate change and habitat destruction, the jerboa is considered of least concern. They are thriving across the multiple countries and continents they call home.

Mola mola

Ocean Sunfish

One of the most unusual-looking fish in the ocean with its flat, circular body, the ocean sunfish can grow to be incredibly large.

There are five species of sunfish found throughout the world's oceans. The ocean sunfish, or *mola* (meaning 'millstone' in Latin), is a remarkable but relatively little-understood animal. This gentle giant is most commonly found swimming on the surface — sunbathing, or 'basking'. This behaviour has given this fish its common name: the sunfish.

The sunfish is the largest vertebrate (spined) fish in the world, and can be up to 3 metres long and 4 metres high from anal fin to dorsal fin. Over time, the sunfish's caudal (or tail) fin has evolved into a thickened area of flesh called the calvus, which the fish uses like a rudder.

Research has suggested that the animal's great size allows it to warm itself rapidly using the sun's rays after it has spent time in deeper, colder waters. Descent into the depths is necessary, as this is where the sunfish finds its primary food source: jellyfish, zooplankton and other fish and crustaceans. The fact that the sunfish is so big and has and thick, rubbery skin means it loses heat relatively slowly. As a result, it can dive deeper and feed for longer.

This enormous fish's 'sunbathing' also has a different purpose. Its body is home to more than 40 species of parasite. As it sunbathes on the surface, smaller fish, such as cleaner wrasse, and birds, such as gulls and albatross, take the opportunity to feed from its skin.

The sunfish has the greatest growth rate of any vertebrate. Females lay up to 300 million eggs each time they spawn. After hatching, the young fish, or larvae, initially just a few millimetres long, grow rapidly. Eventually, they achieve a weight 60 million times heavier than when they hatched.

While not endangered, the sunfish is considered vulnerable. Because of its size, it has few natural predators beyond sharks, orcas and sea lions. The primary danger to it is from humans. Sunfish can easily get caught in fishing nets or choke on plastic pollution in the ocean.

Kallima inachus

Orange Oakleaf Butterfly

It's a leaf. It's clearly a leaf. There is very little to indicate that this isn't a leaf – until it actually takes off and flutters away.

Surprisingly, this 'leaf' is in fact a butterfly. Nature has almost countless examples of astonishingly effective camouflage, but this extraordinary animal is one of my favourites.

Camouflage is an extremely complex subject, and many animals use camouflage for different reasons. It's usually for predators to conceal themselves from their prey or for the prey to avoid being eaten. Mimicry is a form of camouflage where an animal appears to be a completely different animal or object — in the case of this butterfly, a leaf. This animal's fantastic disguise is a great example of how, over huge spans of time, natural selection can adapt an animal in the most incredible ways.

From generation to generation, this butterfly's ancestors became imperceptibly more leaf-like, therefore much less likely to attract the attention of predators such as birds, spiders and wasps. This helped them to survive and to breed increasingly leaf-like offspring.

One of the reasons I find this animal so fascinating is that it doesn't just look like a leaf — it also acts like a leaf. For a butterfly, it is a particularly strong flyer, swishing about among the dense forests of India and South East Asia. However, if it realises that it's being pursued by a bird, it will suddenly close its wings and tumble down onto the leaf-covered forest floor. Here, it is indistinguishable from the other leaves that have fallen to the ground — even as it falls, it will look like a leaf gently falling from a tree. This makes it quite impossible for the pursuing bird to locate it.

Incredibly, this unique butterfly even changes its appearance between the seasons through a phenomenon known as polyphenism, which describes how distinct changes in characteristics can develop in a single species under different environmental conditions. Simply put, this means that the orange oakleaf butterfly displays differences in colour and size depending on the season. In the dry season, they are bigger and lighter, while in the wet season they are smaller and darker.

Although this beautiful and baffling butterfly maintains a stable population, habitat loss presents a potential threat. Its greatest threat, though, comes from the effects of ongoing climate change.

Chlamyphorus truncatus

Pink Fairy Armadillo

Although it isn't actually a fairy, sightings of this adorable armadillo in the wild are very rare as it spends most of its life underground.

The pink fairy armadillo is the smallest of the armadillos, and this tiny inhabitant of the plains of Argentina is wonderfully adapted to its environment. The armadillo is a mainly subterranean creature, meaning that it spends most of its life underground. In fact, it was once thought that it could 'swim' through the sand and soil of its home. It actually burrows, and very rapidly too, using its thick claws as spades, in a similar way to the mole (see page 30).

This armadillo's burrowing lifestyle has caused it to evolve many fascinating traits. It has a streamlined, torpedo-shaped body that allows for easy movement through the ground. Because it spends so much time underground, the armadillo doesn't need good vision, so it has evolved very small eyes. Instead of looking where it is going, it uses its senses of touch, smell and hearing to find its way to its food of grubs, insects and (if pushed) various plants. It has also developed a flattened area of armour, made of keratin like your fingernails, across its rear, which it uses to push back and compress the soil behind it. This helps to

prevent any potentially dangerous cave-ins when it is digging tunnels. Like other armadillos, and also pangolins (see page 50), the pink fairy armadillo is able to curl up into a ball when threatened. This is because its armoured back is hinged and perfectly shaped to protect it.

On the subject of armour, you might be wondering why this armadillo is pink. Its distinctive colour actually comes from the blood vessels which can be seen through its protective armour, or carapace. The colour can change depending on the temperature of the blood and the amount of oxygen within it. The animal also uses this shell as a way to help regulate its temperature, as it can both absorb heat and release it. As is the case with other environments around the world, the weather on the Argentinian plains can change, and can go from very hot to very cold relatively quickly. This little animal's armour is an excellent adaptation for such conditions.

The conservation status of this rare animal is unknown because there isn't enough data available, but it is believed that they are extremely vulnerable. Because the pink fairy armadillo lives in a very specific environment and has evolved to survive within very specific temperature ranges, it is at great risk from climate change. Also, because of its subterranean lifestyle, it is naturally extremely vulnerable to rainfall. Increased rains caused by climate change lead to flooding, which devastates any animal living underground. Further human factors such as farming, increased use of pesticides and hunting are all threatening this animal. Taking these problems into account, it's highly likely that the pink fairy armadillo is very close to the edge of extinction.

Physalia physalis

Portuguese Man-of-War

This sea creature is named after the eighteenth-century man-of-war battleship, as it's considered to look like one at full sail – but its other nickname is 'floating terror'!

Although it may resemble a jellyfish, the man-of-war is in fact a colonial animal called a siphonophore, which is made up of multiple living parts. A siphonophore is a hugely complex creature that starts out as a fertilised egg but, as it develops, begins to branch out or 'bud' into distinct organisms, which form a colony. These individual units that make up the colony are broadly called zooids. All of the zooids are genetically identical, and each one is unable to survive on its own. Over time, these organisms have developed the means to cooperate. Each zooid has a specialism and plays its own part in keeping the animal alive: the pneumatophore is filled with gas and floats on the ocean surface; the gonozooids and gastrozooids are used for reproduction and digestion; and the dactylozooids are used for hunting and capturing prey.

The man-of-war cannot swim on its own, as it doesn't have any active means of propelling itself through the water. Instead, its gas-filled pneumatophore keeps the creature at the water's surface and acts like a sail, allowing it to catch the wind and drift with the tides. When threatened, the man-of-war can deflate this air bag, allowing it to submerge beneath the water for safety.

Man-of-wars are often found in groups of 1,000 or more, drifting in the current, waiting for prey. The man-of-war's tentacles typically grow up to 10 metres in length, with some stretching to 50 metres. These tentacles are extremely venomous, useful for both defence and hunting. Few animals eat the man-of-war, but those that do include the ocean sunfish (see page 62) and sea turtles. The tentacles of the man-of-war can easily detach if they become damaged. Even when removed from the whole colony, a severed tentacle can still deliver an extremely painful sting.

Contrary to what is happening with other species, climate change is actually increasing the population of the man-of-war. The rise in ocean temperatures and reduction in oxygen levels in the water has created an environment where they thrive. In fact, rising ocean temperatures are even likely to result in an increase of tropical sea creatures washing up on beaches in typically colder climates.

Hippocampus bargibanti

Pygmy Seahorse

As its name suggests, the pygmy seahorse is very tiny indeed, typically measuring just a centimetre or two in length.

The Bargibant's pygmy seahorse has evolved to be so wonderfully well camouflaged that it wasn't discovered until the late 1960s. It was only spotted when a scientist who was studying a piece of coral found the tiny animal clinging on to a branch. Today, there are nine known species of pygmy seahorse, seven of which live in the Coral Triangle of South East Asia. In 2018, the *Hippocampus japapigu* (also known as the 'Japan pig') was discovered on the coast of Japan. More recently, in 2020, *Hippocampus nalu*, also known as the African pygmy seahorse, was discovered in South Africa.

From a pygmy seahorse's point of view, its environment is extremely hostile and full of dangerous predators, so it is easier to stay well hidden. This creature lives exclusively on fan corals and has evolved fleshy bumps and protuberances all over its body which perfectly match those of the coral it calls home. It spends the majority of its quiet life holding on to just one piece of coral, and will not move more than a few centimetres away from it for its entire life. Because it is so small, swimming around unnecessarily is a risk it simply cannot afford to take.

As in the case of all types of seahorse, it is the male that incubates its young. The female transfers her eggs over into a special pouch in the male's body, and after a period of around 14 days, the fry are released. In order to survive, they must become instantly independent and seek out the perfect host coral quickly, before they are preyed on. For this reason, this species has evolved to become one of the fastest swimmers in the ocean, relative to body size. Should the need arise, the pygmy seahorse can swim 500 times its own body length per second, which is incredibly speedy considering its tiny size!

It's not known how many of these little creatures exist, as their tiny size and incredibly effective camouflage makes them extremely hard to study, but new species of pygmy seahorse are likely to be discovered even now. Pollution and rising acid levels in the waters they call home are, however, posing a real threat, as they are destroying the corals that they rely on to survive.

Saiga tatarica

Saiga Antelope

It's extremely obvious where the saiga differs from other antelope species: it's the extraordinary bulging nose.

Today, the populations of saiga are limited to relatively small areas of Russia and Kazakhstan. Over time, the saiga's nose has developed to help it survive in these climates, which are often dry, dusty or extremely cold, making them inhospitable to animals. The bulbous and elastic flesh that surrounds its nostrils is packed with blood vessels. This allows the antelope to warm up freezing air or cool down warmer air as it breathes in, as the blood within the vessels adds or removes heat as needed, depending on the conditions. It's also able to filter out the heavy dust that often blows across vast, flat areas of the continent.

The saiga has also developed the ability to change the colour of its fur with the seasons. Its coat will change from a rusty, sandy colour in summer to a paler grey and brown in the winter. This helps to camouflage it from its chief predator, the wolf. Younger animals are also particularly vulnerable to predators such as foxes and eagles.

The male saiga sports a pair of thick and slightly translucent horns. These are, like the horns of other antelope species, for display and defence. A male saiga will fight another male for the 'possession' of a herd of females. These herds can range in size from just a few individuals to as many as 50. This gives an opportunity for the male to reproduce and spread its genes as widely as possible.

The vast saiga herds that used to roam across the majority of Asia are now reduced to a dwindling number. In the late nineteenth and early twentieth centuries, saiga antelopes were severely overhunted for their horns as they were considered to be an alternative to rhinoceros horn, which was incredibly sought in the traditional medicine market.

Sadly, the continuing demand for saiga horn is taking the species ever closer to total extinction, and the saiga is now listed as critically endangered. In the last 15 years, saiga populations have dropped by around 95 per cent, with possibly only 50,000 animals left. Compared with the millions that once existed, these numbers are shocking and extremely worrying in terms of survival. Such seemingly vast numbers of an animal can make the species seem safe – but this often just disguises how vulnerable it is. The saiga is an example of this, along with various species of shark and the American bison (see page 10).

Crocodylus porosus

Saltwater Crocodile

These prehistoric-looking creatures perhaps bring us as close as possible to experiencing what it would have been like to live among dinosaurs.

Over 250 million years ago, prehistoric crocodiles called archosaurs, or 'ruling reptiles', roamed the Earth, long before dinosaurs, birds and modern-day crocodiles. Among the 23 crocodile species of today, the saltwater crocodile is perhaps the most spectacular. They are the largest of the crocodiles and can grow up to a staggering 7 metres in length. They can also live for more than 70 years. Saltwater crocodiles are widespread and can be found in swamps, rivers and brackish water throughout India, South East Asia, Australia and the island of New Guinea.

A fearsome predator, the saltwater crocodile has the strongest bite of any animal on Earth. Its jaw muscles are specially adapted to give an enormous bite pressure, with the aim of gripping and crushing prey rather than chewing it. The jaw may have powerful closing muscles, but the muscles that open it are relatively weak, so it can be held shut easily.

Saltwater crocodiles are not fussy eaters and will take pretty much anything that comes their way. However, they are not greedy and can survive for a long time on relatively small amounts of food. As time goes on and the crocodiles grow to their full size, they can kill larger prey — including, notably, humans. When hunting large animals, the crocodile has developed a technique of rolling its prey. If the animal is particularly large or puts up a fight, the crocodile will pull its prey into the water and instantly begin a rapid rolling motion. This disorientates its prey and works the crocodile's teeth deeper into the flesh. If that's not enough to kill the prey, this rolling motion will eventually drown it.

This crocodile is an ambush predator, meaning it hunts by slowly and patiently sneaking up on its unfortunate prey. It possesses extremely acute senses to help with this. To help it move so stealthily, the crocodile's nostrils, eyes and ears are all located on the upper surface of its head. This allows the animal to be all but invisible as it slowly floats along, with just its eyes and nostrils above the water's surface. It has also evolved extremely sensitive bundles of nerves along its jaw, which are visible as

black speckles. These sensors allow the crocodile to detect the slightest motion in the water over large distances. They are even able to detect a single drop of water hitting the surface. This can give the crocodile a good idea if there is some tasty animal either in or very close to the water.

Most crocodiles aren't particularly threatening and prefer to steer clear of humans. However, the saltwater crocodile is believed to be overly aggressive because of its territorial nature. Researchers who study these animals have shown that crocodiles are very intelligent and relatively social animals. They also have a broad range of vocalisations that they can use to communicate with one another, including growls, hisses and cheeps.

Unfortunately, the saltwater crocodile has been under threat for some time. Thanks to conservation efforts, numbers are bouncing back, but crocodiles are still hunted extensively – occasionally for their meat and skin but more often as trophies for hunters to display. These are some of the most extraordinary animals walking the Earth, perfectly adapted to their environment because of millions of years of evolution, and they need our protection to continue to exist.

Cervus elaphus scoticus

Scottish Red Deer

The Scottish red deer is the largest deer in the United Kingdom, and its impressive antlers make it one of the most recognised species in the world.

Though a familiar sight to most of us, antlers are a strange phenomenon – they are extensions of the animal's skull which protrude above the skin. Only male deer (known as stags) grow antlers, which can reach an impressive size of more than a metre. These protrusions have evolved both as a secondary sexual characteristic to impress females and as a means of defence and attack.

Each year, the red deer will mate in a period known as the rutting season. Around this time, stags will fight each other for territory or for the right to mate with the females of the herd. The successful winners of these fearsome battles show that they will be the most suitable to pass on their genes to create a new generation of strong, healthy offspring. To make themselves seem even larger and more impressive, males sometimes roll around on the ground, collecting vegetation in their antlers in a process called adorning.

At the end of each rutting season, the stags will shed their antlers and almost immediately begin to grow new ones. They are grown from 'buds' within the deer's skull and can grow at a rate of up to 6 centimetres every day, depending on the nutrients the deer gets from the available food. This makes them one of the fastest-growing 'horns' of any animal on the planet.

At first, these antlers are relatively soft, with a smooth covering called 'velvet', which is a bit like human skin. This protects the new antlers and is also packed with blood vessels that provide nutrients to the bone as it grows. Once the antlers have finished growing, the velvet covering sheds, drying out and peeling away. This process can look quite startling, as the velvet hangs off the antlers in long, bloody tatters, but it is painless to the deer. In fact, stags can often be seen rubbing their antlers against trees attempting to remove these bothersome shreds as quickly as possible.

Apart from impressive antlers, there is another extremely fascinating thing to be noted about this deer. A recent 45-year study focusing on the red deer population on the Isle of Rum in the Inner Hebrides found that female deer (known as does) are now giving birth on average 12 days earlier than they were in the 1980s. Further investigation found that this was a direct effect of climate change.

Balaeniceps rex

Shoebill

When you first set eyes on a shoebill, you could be forgiven for thinking you were looking at a dinosaur.

In some ways, the shoebill's prehistoric appearance derives from the fact that birds are closely related to dinosaurs. In broad terms, a grouping of dinosaurs called theropods, which included the Tyrannosaurus rex, evolved into modern birds.

Living in the tropical swamps of eastern Africa presents a whole host of challenges that have all influenced the shoebill's evolution. These swamps are dangerous places. Predators are around every corner, disease is rampant, the heat can be extreme and food is often scarce. In this hostile environment, it pays to be somewhat hostile yourself, and the shoebill is a particularly aggressive animal.

From birth, the young birds will instinctively fight one another for dominance of the nest and the attention of their parents. Often, the siblings can start behaving so aggressively that some chicks will fly the nest to survive, often before they are ready. The chances of survival are greater when they are alone than they would be if they continued to live with their siblings. It is a perfect example of the evolutionary principal of 'survival of the fittest'.

The shoebill has become a formidable and fearless hunter, which is a great benefit when food can be a limited resource. Hunting mainly at night, the shoebill will wade through the shadows of the swamp with its long legs and large webbed feet helping it move through the water and long grass. It then stands motionless, patiently waiting for a suitable meal to pass by, then lunging to attack at great speed. Fish or birds might fall foul of this trap — even young crocodiles.

The shoebill's most prominent feature is its huge, strong beak, which is usually about a quarter of the size of its entire body. The beak has a hooked end designed for spearing its prey and for gripping it securely to stop it from moving. The shoebill also uses its beak for digging in the soft mud of the swamp in search of lungfish and other creatures that dwell in the depths to feed to its young.

As a territorial animal, the shoebill will use its beak to make a loud clacking sound, as either a mating or a warning call. Instead of using the syrinx (the equivalent of our voice box), like other birds do, the shoebill claps its beak together to make an intimidating sound that echoes across the swamp.

The species is listed as vulnerable, with populations numbering around 5,000–8,000 animals. The primary threat to the shoebill is habitat destruction, with areas of swamp frequently cleared for agriculture. The bird is often hunted or captured for private collectors, and its eggs are also frequently taken for food. One can only hope that a viable population of these stunning birds survives for the future.

Casuarius casuarius

Southern Cassowary

These birds are little understood. They are so shy that scientists find it extremely hard to study them in the wild.

The southern cassowary is a large bird, standing up to 1.7 metres tall. They are also expertly camouflaged and blend in perfectly with their environment of the rainforests of Australia and New Guinea. To help them live in these forests, they have developed various adaptations — one of the most noticeable is their shape. Cassowaries are flightless birds, and their wing feathers have been reduced to essentially a few rigid quills. These feathers are tightly packed and, together with their heavy, wedge-shaped body, allow them to push rapidly through the dense undergrowth of their home without getting hurt by sharp leaves and thorns. And I really do mean rapidly — they can run at a speed of up to 50 kilometres per hour!

The cassowary's feet are another useful feature. They are three-toed, with the middle toe armed with a fearsome claw, making for an extremely efficient form of defence. When they feel threatened, cassowaries can kick out with this claw, and they have been known to kill humans and other animals easily. However, this behaviour is entirely defensive. Although the cassowary is omnivorous and occasionally eats small animals, its diet generally consists of fruit and seeds.

This animal's most striking feature is the helmet-like horn on the top of its head. It is known as a casque, and there are several theories for its use, none of which are definite. It's possibly used for mating displays, along with behaviours such as strutting and headshaking. It has also been suggested that the casque might act as a horn, in a similar way to a rhino's horn. The cassowary can use this casque to push its way through the dense rainforest, to shift aside earth and leaf litter when searching for food, and as a defence when fighting. One of the casque's most intriguing functions is that it works as an amplifier. To communicate efficiently in the dense jungle, the cassowary produces a very deep, low-frequency booming call, almost inaudible to human ears. The casque seems to help amplify and deepen that sound, allowing it to travel further.

The southern cassowary of Australia is listed as endangered, mainly because of habitat loss. As a result of significant rainforest destruction, where forest is being cleared for cattle production, only an estimated 25 per cent of its original habitat remains.

Physeter macrocephalus

Sperm Whale

*T**he whale has been the stuff of myth and legend for as long as humans have ventured across the world's oceans. Looking at the sperm whale, it's easy to see why.*

The sperm whale is truly awe-inspiring. It is are the largest toothed predator in the world, measures around 15 metres or more in length, and weighs around 40 metric tons. Some have been recorded as being much larger, but these giant whales are very rare. The sperm whale has several unique adaptations that single it out from other whale species.

It has large, conical teeth that allow it to efficiently grab its prey of squid and octopus — also, more dangerously, giant squid. The titanic battles between these two giant sea-dwellers often scar the whale's skin with deep grooves and welts from the squid's suckers.

The sperm whale also possesses the largest brain on the planet and is intelligent and extremely social. In addition, the whale uses a echolocation ability.

This echolocation, which you may hear referred to as sonar, allows the whale to produce sounds, such as clicks and whistles, that travel through the water, strike an object, and are reflected back to the whale. It then hears these reflected sounds and uses the information to locate the object the sound bounced off. Some animals use sonar in the air rather than in water — bats, for example (see page 86). To produce its sonar sounds, the sperm whale has phonic lips in its nose. As air pushes through the nasal passage and past the phonic lips, the surrounding tissue vibrates, producing the loudest sounds of any animal.

The sperm whale's incredible head, which takes up almost a third of its body, is almost entirely devoted to sonar. Within its skull is a huge cavity containing an oily liquid called spermaceti. This, along with other complex fluid-filled structures in the whale's body, works as both an amplifier and a reflector for the sounds the whale produces.

There are fluid-filled structures beneath a whale's echolocation-producing organ that act as cushions for the whale when it rams into objects or other animals. This strategy was famously shown in the real-life incident of the whaling ship *Essex*, which inspired Herman Melville's novel *Moby-Dick*.

For a long period of time, spermaceti, a highly flammable substance, was used to fuel lamps. The harvesting of this oil led to the hunting of sperm whales in large numbers, causing many populations to remain at risk even after whale hunting was stopped in the late 1980s. Unfortunately, some countries continue to engage in this practice. Whales are also threatened by climate change and pollution, as well as becoming entangled in commercial fishing nets.

Ateles

Spider Monkey

Studies have indicated that the spider monkey is one of the most intelligent monkeys found in the Americas.

Spider monkeys have developed many features to help them in their arboreal (tree-dwelling) existence. Firstly, their hands have changed over time to essentially work as hooks. Their thumbs have pretty much disappeared, and their long fingers are narrow and curved. Along with their extremely long arms, this adaptation gives them extraordinary agility among the branches of the canopy.

In addition to this, its tail has evolved into a fifth limb, which is prehensile, meaning it can grasp onto objects. Over time, this tail has developed a mobile tip, which can easily wrap around branches or tree trunks. This tip is also hairless, with a ridged and grooved surface that gives the monkey tremendous grip on branches.

Spider monkeys are extremely intelligent, with a large brain. This feature may have developed — in part — because of their diet. These monkeys live on a wide variety of fruits, nuts and berries, many of which ripen at different times of year and in different areas of their territory. The monkeys need to be able to remember which food grows where, and when it can be eaten, which requires a lot of brain power.

As primates, spider monkeys also have very complex social systems and remarkable vocalisations known as whinnies, which help them locate and identify one another. They are long-lived, with a lifespan of 25 years or more (particularly in captivity). All of this has allowed their brains to become more developed compared with the brains of similar animals, such as the howler monkey.

The spider monkey, found in the dense forests of the northern Amazon in Ecuador, Peru and Brazil, is now sadly in great danger from the deforestation of its habitat, with the white-bellied spider monkey now an endangered species. In the last 40 years or so, spider monkey numbers have been reduced by half. The trees they call home are being destroyed at a truly terrifying rate. They are also widely hunted for their meat and, unfortunately, they're used extensively in medical research. Such a beautiful and unique animal deserves to be protected — as does all of nature.

Euderma maculatum

Spotted Bat

Bats are sometimes associated with darkness and death, but nothing is more beautiful than seeing these incredible creatures fly through the night.

There are more than 1,400 species of bat in the world, all quite different and living in various places. The spotted bat is found in western areas of North America, such as California, Arizona and parts of Mexico. If you live in or visit these areas, keep an ear out! The echolocation sounds that bats make are normally out of the range of human hearing, but the spotted bat is one species you might be able to hear.

Echolocation is what a bat uses to hunt in the dark — similar to the echolocation of dolphins and whales (see page 82). The bat emits a rapid series of clicks; these sounds bounce off an object, perhaps a tree branch to be avoided or a juicy moth to be caught. The sounds reflect back to the bat, where they are picked up by its extremely sensitive (and in this case, enormous) ears.

The spotted bat's ears are unique among bat species as they are so huge. They are also very delicate and easily damaged — problematic, as a bat without perfectly functioning ears essentially has a death sentence. But the spotted bat has a strategy to avoid any harm coming to its ears. It has evolved the ability to fold them away across its back when it is resting. When it's ready to hunt again, the bat's ears refill with blood and return to their normal, glorious size.

Bat species vary wildly in size — some are as small as a bumblebee! The spotted bat is of average size, about 10 centimetres long. With a certain amount of luck, most bat species can live for around 20 years in the wild. Little is known about the spotted bat, but its lifespan is assumed to be within this range.

Unusually for bats, the spotted bat is solitary and lives in high roosts in desert regions. It has excellent hearing, can use echolocation expertly, and is also a very nimble and fast flyer. This all means that the spotted bat has little to fear from predators. Very little is known about population numbers for this species, but recent observations suggest that the numbers are, thankfully, increasing.

Crocuta crocuta

Spotted Hyena

Hyenas have long been considered devious and ugly scavengers. The irony, of course, is that they are the exact opposite of these stereotypes. They are extremely intelligent and beautiful animals.

Although they are often thought to be scavengers that feed on the prey of other predators, studies have shown that spotted hyenas kill up to 90 per cent of the animals they eat. They have an extremely broad diet, which ranges from small mammals to buffalo and giraffe. Their great intelligence makes them effective hunters, as they are able to work as a pack, plan ahead and patiently wait for the ideal circumstances for the most successful kill.

These incredible predators have evolved remarkably strong jaws, containing powerful muscles. Not only can hyenas' teeth pierce thick hides, they can also crack the bones of their prey. This allows them to access the nutritious marrow within the bones, which is inaccessible to other animals. They have also evolved a strong digestive tract that can break down almost the entirety of the animals they hunt — meat, skin, fur and bones.

Hyenas are also aided by incredibly acute senses. They have excellent sight and smell, and their hearing is so sharp that they can pick up the sounds of another predator's kill from around 8 kilometres away. Although spotted hyenas do not scavenge much, they will not pass up the chance of a good meal.

Contrary to popular belief, spotted hyenas are in fact intelligent, curious, inquisitive and highly social. There is strong evidence that the evolution of hyena intelligence closely mirrors our own, with complex social skills driving the evolution of the brain. Experiments have indicated that hyenas are better problem-solvers than many of the great apes, both as individuals and in groups. They can also count!

Hyena packs, called clans, can number up to 100 individuals, and are led by females. Female hyenas are fiercely protective of their cubs — they will not allow any other adult hyena to approach them as they're maturing. Once they have reached adulthood, males will leave their own clans and seek another group to join. Females always have seniority over males in any clan, and male hierarchy is determined by who has been in the clan the longest. When a male joins a new clan, he will gauge the number of other males in the clan to determine his rank.

Although the spotted hyena is considered of least concern from a conservation point of view, their situation must be watched carefully. They are in particular danger in areas where humans encroach upon their territory for new agricultural land. Hyena clans will sometimes begin hunting livestock and are therefore often killed on sight. Although their age-old reputation precedes them, I hope that in the future these intelligent animals will be looked at in a new light.

Mephitis mephitis

Striped Skunk

Feared by humans owing to the pungent odour they emit when threatened, skunks have been represented as a misfit in both popular culture and myth, but these beautiful and peaceful creatures have long been misunderstood.

Skunks are small animals, about the size of a house cat, that live across North America. They are often found in burrows or hollow logs. Skunk fur comes in a variety of patterns, including stripes, spots or swirls — but one thing that all of these patterns have in common is that, to other animals, they spell out, 'Keep away!'

Any animal that messes with a skunk really will regret it. It has special glands near its bottom which are filled with a substance called musk, which it can spray when provoked, giving off one of the most revolting and putrid smells you could ever encounter. This spray has evolved entirely as a defence mechanism.

The skunk is a timid, nocturnal animal that is no danger to anything other than its prey. It mostly hunts insects, small reptiles and fish, and also supplements its diet with fruit and seeds. When it does feel alarmed, the skunk will stomp its feet in warning and perform a 'handstand' on its front paws before it prepares to spray. If it continues to feel threatened, the skunk will return to all fours and turn its bottom towards the threat.

Although various cartoons might make you think that a skunk can go round spraying things at the slightest provocation, skunks have to produce their spray in specialised glands and can't instantly 'reload'. Skunks have to feel particularly threatened and take careful aim before they unleash their spray.

The spray is certainly effective — it means the skunk doesn't have a great many natural predators. Despite its temptation as a meal, predators tend to keep their distance. This is both instinctive, because of the warning patterns on the skunk's fur, and learned from experience. Some predators, such as cougars, foxes and eagles, will attempt to take on a skunk, but generally only when they are absolutely starving — a 'nothing to lose' philosophy.

Despite being killed for its meat and fur, and, at one point, for medicinal purposes, this delightful, charming animal isn't considered endangered. Some people even keep them as pets.

Dicerorhinus sumatrensis

Sumatran Rhino

A beautiful and solitary animal, the Sumatran rhino is the smallest and by far the most threatened of the world's rhino species.

The Sumatran rhinoceros is one of five existing species of rhino and is perhaps the most unusual. Living in the dense forests of Sumatra, an island in western Indonesia, this rhino has evolved several unique adaptations to its environment.

Unlike its cousins (the more familiar black, white, Indian and Javan rhinoceroses), the Sumatran rhino is covered in a coat of shaggy, reddish-brown hair. There are many reasons why the rhino might have this hair — it might be to regulate its body temperature, as this animal has very little subcutaneous fat (the layer of fat just underneath the surface of the skin). The presence of this hairy coat has also led researchers to believe that the Sumatran rhino is closely related to the woolly rhinoceros, an animal that became extinct after the last ice age about 1,000 years ago.

Its hair is not the only feature that sets this rhino apart from some of its relatives. Like the black and white rhinos (the two African species), the Sumatran rhino has two horns, although they are typically much smaller, and the rear horn is just a small stump. The reasons for this are not clearly understood. However, a possible reason lies in these animals' solitary nature — horns can be used for signalling to other rhinos, but the Sumatran rhino has no need to do this. These animals also have no natural predators, so no longer need large horns for the purposes of defence. In addition, a large horn could be a hindrance in moving freely around their thick forest environment, with its dense trees and vines. Instead, the horns seem to be used mainly for foraging and scraping up the ground in search of roots to eat.

Sadly, this gentle animal is extremely endangered. The rhino is hunted for its horn, which is seen to be valuable in certain cultures. Habitat loss as a result of deforestation and agriculture has also pushed the rhino into smaller forested areas, leaving them even more vulnerable to poachers. Conservation efforts are in place to help tackle the illegal wildlife trade and protect the rhino's habitat. One can only hope that this beautiful, unique animal survives.

Galeopterus variegatus

Sunda Colugo

The Sunda colugo is also known as the Malayan flying lemur, but it is neither a lemur nor capable of true flight. Rather than flying by powering itself through the air, it simply glides.

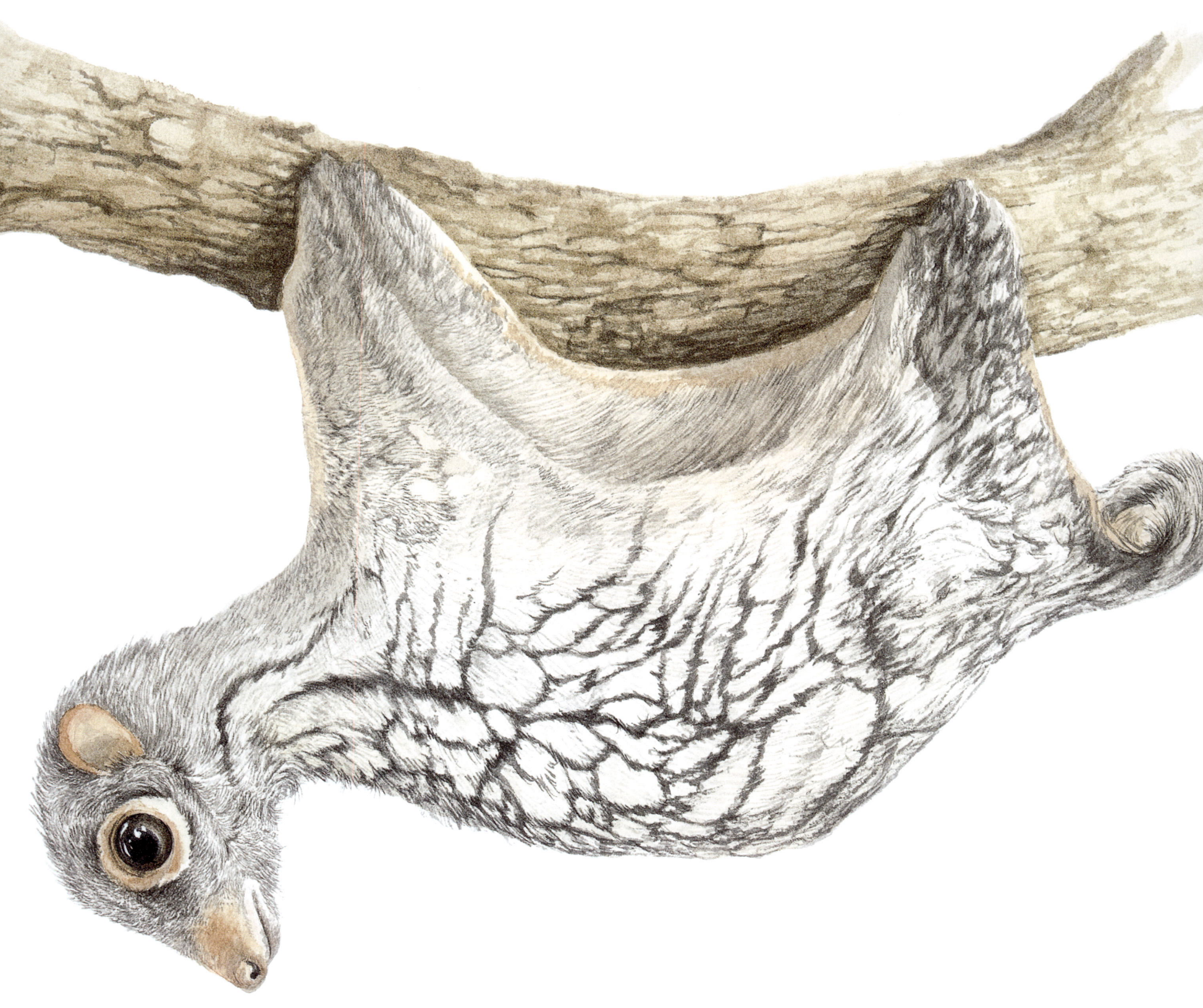

The colugo has evolved a membrane of skin that stretches out from its neck, connecting its arms, legs, fingers and toes. This effectively turns the animal into a kite and allows it to glide for up to 100 metres in one leap. This is clearly an enormous advantage to its survival, as it allows the colugo to avoid any potential threats swiftly by simply launching itself from tree to tree. It also means the colugo is rapidly able to locate food sources that could be scarce.

Found widely in the jungles throughout Indonesia, the colugo has become so wonderfully adapted to its environment that it lives almost entirely within the forest canopy. The adaptations that let it glide have made it essentially helpless on the forest floor. In fact, its ability to glide becomes a significant hindrance when it is out of the treetops as the gliding membranes get in the way when trying to negotiate the complex forest floor.

It is fascinating to imagine how the colugo's kite-like membrane could have evolved. Picture a small mammal, living a long time ago, seeking food and safety in the forest canopy. It takes a small leap, perhaps to avoid a predator, from one low branch to another. This particular animal has a slightly wider web of skin between its arms and legs than normal. The leap is successful – it escapes! It has survived, so can go on to find a mate. Its offspring inherit this small advantage, and they too successfully mate and produce their own offspring. Some of these future colugos have slightly wider webs of skin than previous generations – one perhaps has a small extra membrane between its neck and arm. Their survival chances are increased. More offspring are produced, and the cycle continues. Uncountable generations of colugo are born and die, but each animal that successfully produces young passes on useful adaptations over time. The colugo's flaps of skin have grown bigger and bigger, creating the colugo species as we know it today. Now, they flourish high in the canopy, launching themselves from tree to tree, their arms and legs outstretched, catching the warm air.

It should be said that this adaptation isn't without its risks. The colugo needs each jump to be over a long enough distance for it to effectively and safely catch enough air to make a successful 'flight'. If a jump is too short, it risks a potentially dangerous landing. However, for the colugo, the risks are worth it. Food can be scarce, even in the jungle, and predators have ingenious adaptations of their own. The benefits of being able to glide great distances outweigh any of the risks.

These gentle animals are thankfully considered of 'least concern' in terms of numbers. Despite this, they are hunted by local peoples and are suffering because of habitat destruction. This is caused by illegal logging and deforestation for agriculture. At least for the present, they are numerous enough to be surviving well despite these challenges that humans are introducing to their peaceful lives.

Bufonidae

Toad

Many people find toads particularly unappealing. However, once you know a little about them, I would hope that you'd find them as beautiful as I do.

Toads come in a vast array of sizes and colours. They are found all over the world in a huge range of habitats, even in extreme places such as deserts. There are about 350 species currently known. Scientifically, there is no distinction between frogs and toads. However, there is a cultural distinction between them, which is referred to as folk taxonomy. Toads are associated with rougher, tougher and drier skin that has bumps and nodules, commonly (and incorrectly) referred to as warts.

Toads possess some remarkable traits, both for hunting and for their own protection. They are not animals blessed with the ability to quickly chase down their prey, but instead, they have developed a long, sticky tongue that they can launch at their prey with enormous speed. In a similar way to the chameleon (see page 22), a toad is able to remain relatively stationary while it waits for some suitably zesty morsel to come wandering past, such as a slug, worm, or insect. When a toad sees something appealing, it can rapidly launch out its tongue, grasp its prey with the sticky end and pull it into its mouth – all quite literally in the blink of an eye.

On the subject of eyes, toads have to blink to swallow. To move the food down their throat, they literally have to push their eyes down onto the roof of their mouth. In all honesty, this isn't the most appealing sight you will ever see – but it is a remarkable adaptation.

Many toads can also produce poison, usually a burning, milky substance that is released from glands behind their eyes. Many animals have learned that this makes a toad an unappealing meal. However, natural selection is nothing if not a fair fight, and many of the toad's predators, such as certain species of snake, have become immune to their toxin.

Toads are amphibians, meaning they live both in water and on land. During mating season, toads will gather near water and call for mates. After mating, the female will lay many eggs within the water, which the male will then fertilise. The eggs and spawn are a tasty snack for many creatures, but those that survive will begin to develop, changing shape in a sequence of distinct forms. These changes are fascinating – watching a tadpole grow is almost like watching a speeded-up process of evolution. The young toad starts out as essentially a small, round blob with a tail. Over time, the tadpole quickly develops legs and hands with distinct fingers. Next, the tail is absorbed into its body, lungs develop, the mouth widens, and its intestines change to allow it to have an animal-based diet rather than just eating plants. Eventually, the tadpole becomes a perfectly miniaturised toad with a potentially long life ahead of it: toads can sometimes live for 15 years in the wild.

Toads also serve a broader ecological 'purpose', as they are both predator and prey. They hunt invertebrates, thereby keeping potentially destructive numbers down, helping to protect crops. Toads are also a food source themselves for birds, other amphibians reptiles and mammals.

Some species of toad are endangered owing to habitat loss, but broadly speaking, they are extremely common, as they have adapted to different environments. Should you be lucky enough to see one, I hope you look at it with the wonder it deserves.

Membracidae

Treehopper

These incredible insects look like something from another planet. This treehopper from the Amazon rainforest of Ecuador is from the genus Cladonota *and is an excellent example of the species' unusual appearance.*

There are about 3,200 species of treehoppers and thorn bugs, or membracidae, in the world, and they are found on every continent except Antarctica, with only five known species in Europe. Treehoppers vary greatly in appearance, ranging in colour from green and blue to brown and often marked with stripes or spots. Treehoppers are best known for their enlarged and ornate pronotum, the plate-like structure (often called a 'helmet') that covers the thorax and expands up and over their back. The pronotum is sometimes formed into incredible shapes that enhance the treehopper's camouflage; they often resemble plant thorns, which is why several treehopper species are called thorn bugs. These varied and impressive appearances help treehoppers evade or scare off predators as well as attract a mate. This creature is undeniably one of the most stunning insects.

Despite its small size, which ranges from about 2 millimetres to 2 centimetres in length, the treehopper has a powerful body and will fly off when disturbed, using specialised muscles in its hind legs that unfurl to create an enormous jumping force. To communicate, treehoppers shake their body to create vibrations, alerting one another of predators or calling for a mate. Each species has a different vibration pattern, which goes through the branch on which its perched and transmits as far as 1 metre away.

Treehoppers poke holes in tree branches and use their tubular mouthparts to suck up the sugary sap inside. They have a special saliva they inject into the hole to prevent a tree from closing up the punctured area, allowing them to keep feeding. Certain species of treehopper secrete honeydew, which is a sugary, sticky liquid made from excess sap during digestion. The honeydew is often consumed by ants, which 'farm' the treehoppers by stroking them with their antennae to cause droplets of honeydew to release from the treehopper's abdomen. This relationship is mutually beneficial for both insects, as the ants provide protection for the treehoppers in return for food. The ants also prevent the excess honeydew from damaging trees and plants, as it could cause mould to grow on leaves and fruit.

Today, scientists are still discovering new species of treehopper, with one species being named after the pop singer Lady Gaga. *Kaikaia gaga* is dark purple and red in colour with two horns jutting out of the top of its helmet, which have been compared to shoulder pads. Who knows what the next discovery in this fascinating family of bugs will be named!

Vampyroteuthis infernalis

Vampire Squid

The vampire squid's scientific name, Vampyroteuthis infernalis, literally means 'the vampire squid from hell'.

Although it is named after the predatory mythical creature the vampire, this squid does not feed on blood. Its diet is made up of detritus and dead plankton that float down from the waters above. That being said, it is easy to see how it could be called a vampire. Its eight arms, connected by a fleshy cape, are each lined with a double row of ferocious-looking spines, adapted for defence rather than attack. Depending on the lighting conditions, the vampire squid appears reddish to almost pitch-black in colour, with piercing blue eyes. In fact, compared to its body size, the squid has the largest eyes in the animal kingdom. This is a useful adaptation for its deep-sea habitat, where there is very little light.

Living incredibly deep in the ocean, around 600–900 metres beneath the surface, the vampire squid has large gills that allow it to absorb as much oxygen as possible from the water. It also digests its food very slowly and has a low metabolic rate, so it uses up less of the energy it gets from its food.

The vampire squid's body weighs very little and contains statoliths (tiny 'stones' made of calcium), which make it extremely buoyant. To help keep it from the mercy of the ocean currents, this squid has specialised organs that help it orientate itself in the water, as well as two short fins that look almost like ears, which it uses to propel itself.

Like other deep-sea creatures, the vampire squid possesses photophores, which are light-producing organs. The highest concentration of these organs is at the base of the fins and the tips of the arms. The squid can use photophores to produce light to confuse other animals.

Unlike other squid and octopuses, the vampire squid cannot produce clouds of ink. Instead, it does something even more remarkable: it expels a cloud of sticky, jelly-like mucus, containing orbs of blue light that can glow for up to 10 minutes, disorientating would-be predators and lighting them up in the darkness so that the vampire squid can escape.

The population size of vampire squid is currently unknown as their environment makes them extremely hard to study – there is no light, the temperatures are incredibly cold and the pressure is immense. However, vampire squid are not yet considered to be threatened. Instead of fearing them, we should admire this incredible creature from the ocean depths.

Giraffa camelopardalis peralta

West African Giraffe

Surely an animal that needs no introduction, the giraffe must be one of the most recognisable animals on Earth. There really is nothing like it.

Firstly, the giraffe is the tallest animal on Earth. When fully grown, an adult giraffe can reach almost 6 metres tall. Measuring about 2 metres, their legs alone are taller than an average person. These long legs are an extremely efficient means of travelling around, allowing them to run at a speed of more than 55 kilometres per hour. However, they are also weapons. When threatened by a predator, such as a lion, the giraffe is able to kick out with great speed and strength. The lion faces a very real possibility of severe injury.

No discussion of the giraffe would be complete without a mention of the animal's extraordinary neck. When baby giraffes are born, their necks are proportionally shorter than those of adults, which makes birthing easier. The giraffe's neck grows as the animal matures, and when fully grown, it can be up to 2 metres long.

The giraffe's neck is a perfect example of evolution giving an animal an advantage over its competitors. Many animals in the giraffe's home of sub-Saharan Africa share their diet — for example, the gazelle and other species of antelope. This competition has potentially encouraged the giraffe's neck to become longer over time, enabling the giraffe to reach food that other animals can't. During the mating season, male giraffes will use their necks as weapons. The animal that remains the most upright (indicating the strongest neck) is deemed successful. The animals who win these duels have been shown to enjoy greater mating opportunities than their rivals. Male giraffes also have two small horns on their head called ossicones, which are specially adapted for fighting. Though these short, hair-covered horns are blunt, they can inflict damage when giraffes come to blows.

The skin of a giraffe is patterned with spots, which vary in colour and shape among the four distinct species, as well as between individual giraffes. In fact, they are as unique as our own fingerprints. These spots are effective camouflage for giraffes. As they spend a great deal of their time feeding among the trees, their spots have evolved to mimic the dappled light that comes through the treetops, sometimes making them all but invisible.

Of the four giraffe species, two are listed as critically endangered, one as endangered and the other as vulnerable. Although great efforts are being made to stop numbers from falling, giraffes are under threat from habitat loss and the bush meat market. But there are some success stories — with community and education programmes along with the protection of national parks, the future is hopeful for the giraffe.

Pithecia pithecia

White-Faced Saki Monkey

The striking colour of its ruff makes the male of this species one of the most handsome primates.

The white-faced saki, also known as the Guianan saki, and sometimes the golden-faced saki, is a small South American species of monkey found in Brazil, Guyana, Venezuela and French Guiana.

The magnificent male saki has a prominent and unique golden-yellow ruff surrounding its face, making it instantly identifiable. This amazing ruff is considered a secondary sexual characteristic – the males have developed this feature for display to other monkeys, and to let females know that they would make a good mate. While this striking ruff may look odd to us, it is incredibly irresistible to the females, and will influence their choice of mate. Therefore, the more impressive the ruff, the higher the chances are of securing a partner.

Both males and females are very similar at birth and are hard to tell apart for the first few years of life. However, as they age, the males develop this pronounced yellow ruff, while the females grow shorter hair and develop two vertical lines of white hair running from their eyes to their mouths.

Like most monkeys, sakis live in small family groups. In captivity, pairs typically mate for life, but scientists are beginning to discover that this is not necessarily true for these animals in the wild. To communicate with each other, they have a selection of distinctive calls. A loud screech or bird-like call is used to attract other family members, while a ferocious roar helps to defend their territory from other family groups and predators.

As they spend almost their whole life in the trees of the jungle, these monkeys are equipped with strong legs that allow them to make huge leaps from branch to branch, reaching distances of up to 10 metres in just one jump.

Thankfully, these beautiful creatures are not currently classified as endangered, living primarily in their natural habitat found within protected areas in Brazil. However, it is difficult to estimate a precise population in the wild, and white-faced saki monkeys are sadly under threat from deforestation. Unfortunately, they are also hunted for their meat and their highly valued tails, as well as being victims of the illegal wildlife trade, often being captured from the wild and sold as pets.

Sus scrofa

Wild Boar

The wild boar is one of the most widely distributed mammals in the world. It can be found on almost every continent, except for Antarctica, either natively or from being introduced by humans.

The wild boar is a successful animal — this is due to its ability to adapt quickly to a huge range of environments. The main reason for its adaptability is that it can eat an incredibly wide variety of foods. The wild boar is an omnivore, meaning its diet includes both plants and animals. Although the roots, bulbs and bark it feeds on are available all year round, the time of availability of each depends on the plant. It also eats earthworms, insects, molluscs, fish, rodents, birds and bird eggs, lizards, snakes, frogs and carrion (dead animals).

To help them find food, boars have evolved an exceptionally acute sense of smell, using their long snout to sniff out prey. They also have very sensitive hearing, which helps them both to hunt and to evade predators.

The boar's head makes up approximately a third of its body length. It has huge muscled shoulders, which enable it to use its head like a plough, pushing deep into the ground in the search for food — even when the ground is frozen.

During the breeding season, female wild boars will typically birth four to six babies at a time. However, if need be, they can reproduce all year round, an adaptation that helps them to survive. Of course, some of their young will be lost to predators — such as wolves, bears and humans — but not so many that populations become threatened.

Boars are also noted for their very prominent canine teeth. These are more pronounced in males and are mostly used for fighting during mating seasons. They can grow to around 5–10 centimetres in length and are very sharp, so they can cause severe, sometimes fatal, damage. In fact, during mating seasons, the males will develop a thick layer of tissue across their shoulders that gives them extra protection should they come to blows.

Owing to their great numbers and ability to survive in such a wide range of habitats, the wild boar is not considered at risk. However, in some areas, the boars themselves have become the threat. Humans have introduced them to ecosystems that are not used to such animals, inadvertently destroying the habitats of native species.

Glossary

Adaptation The process of change that means an animal is better suited to its environment.

Ancestor A relative who lived a long time ago.

Camouflage Colours or patterns that help an animal to blend in with its surroundings and hide from predators.

Carnivore An animal that eats meat.

Climate change A large-scale and long-term shift in Earth's weather patterns and average temperatures; effects include rising temperatures and changing weather patterns.

Cold-blooded Having a body temperature that changes according to the environment.

Conservation The act of protecting ecosystems and environments to protect the animals that live there.

Deforestation The clearing or cutting down of forests.

Endangered Describes a species at risk of becoming extinct.

Evolution The process by which living things change over time.

Extinct Describes a species of animal or plant that no longer has living members.

Habitat The places where plants and animals live. A habitat provides the right type of food, water and shelter for the plants and animals that live there.

Herbivore An animal that eats plants.

Insectivore An animal or plant that eats insects.

Insulation A thick layer that stops heat escaping.

Metabolism The chemical reactions in the body's cells that change food into energy.

Migration The movement of animals from one region to another at a particular time of year.

Nocturnal Awake and active at night, and asleep during the day.

Omnivore An animal that eats both meat and plants.

Plumage All the feathers on a bird's body.

Poaching The illegal hunting of animals.

Predator An animal that hunts other animals for food.

Prehistoric The period of history before anything was written down.

Prey An animal that is hunted or killed by another for food.

Savannah A large flat area of land covered with grass. Usually containing very few trees, Savannahs cover half of Africa and large areas of South America, Australia and India.

Spawn Animals that live in water such as fish and amphibians lay their eggs, or 'spawn', in water. This type of reproduction is called 'spawning'.

Species A particular type of plant, animal or other living thing.

Toxin A poison produced by a living thing.

Territory An area that is occupied and defended by an animal or group of animals.

Venomous Able to produce a fluid called venom which is poisonous to animals and humans.

Warm-blooded Having a body temperature that usually remains the same. Warm-blooded animals can make their own heat, even when it is cold outside.

Wetlands An area of land that is covered by water, either permanently or at certain times of the year. Swamps and marshes are examples of wetlands.

About the Artist

With this book, William Spring returns to his love of natural history painting with the philosophy that there is nothing in nature that is not beautiful. Spring began his career as a newspaper and magazine cartoonist and has also worked as an archaeological illustrator. He has written and illustrated more than a dozen picture books for children under the pseudonym Alborozo. William Spring lives in England.

A Note from the Artist

During the making of this book, I began to lose the sight in one eye. Apart from the obvious challenges this brought for me as an artist, it also highlighted the point of this book, that every single living thing on this planet is beautiful, and to witness it, to see it for what it is and represents, is truly a wonder. Each and every example of life represents an unbroken line of survival and evolution spanning billions of years. It is our duty to protect and nurture it. We cannot survive without it.